# Series in BioEngineering

**Series Editor**

Magdalena Stoeva, Medical Imaging Department
Medical University Plovdiv
Plovdiv, Bulgaria

The Series in Bioengineering serves as an information source for a professional audience in science and technology as well as for advanced students. It covers all applications of the physical sciences and technology to medicine and the life sciences. Its scope ranges from bioengineering, biomedical and clinical engineering to biophysics, biomechanics, biomaterials, and bioinformatics.

Indexed by WTI Frankfurt eG, zbMATH.

Almir Badnjević • Lemana Spahić

# Intelligent Systems in Biomedicine

## Theories, Algorithms, and Clinical Applications

Springer

Almir Badnjević
Registers and Data Exchange of B & H
Agency for Identification Documents
Banja Luka, Bosnia and Herzegovina

Lemana Spahić
Research Institute Verlab
Sarajevo, Bosnia and Herzegovina

ISSN 2196-8861 ISSN 2196-887X (electronic)
Series in BioEngineering
ISBN 978-3-032-14784-4 ISBN 978-3-032-14785-1 (eBook)
https://doi.org/10.1007/978-3-032-14785-1



This Springer imprint is published by the registered company Springer Nature Switzerland AG
The registered company address is: Gewerbestrasse 11, 6330 Cham, Switzerland

# Foreword

The study of intelligent systems has evolved from a visionary concept into a cornerstone of modern science and technology. Today, artificial intelligence and machine learning are no longer confined to theoretical discussions; they are integral to advancing healthcare, industry, and society at large. Few fields demonstrate this transformative power as clearly as biomedicine, where intelligent systems are helping to diagnose disease, guide treatment, and personalize patient care in ways that were unimaginable only a generation ago.

This textbook arrives at a critical moment in that evolution. It provides a rigorous yet accessible foundation in intelligent systems while situating these concepts within one of the most impactful domains of application: biomedical engineering and healthcare. The reader is taken on a carefully structured journey, from the origins of computational intelligence and the fundamentals of neural networks, to cutting-edge developments in generative models, transformers, and ethical frameworks for responsible AI.

What distinguishes this work is its breadth and depth. It does not treat intelligent systems merely as abstract algorithms, but as tools with profound real-world consequences. By weaving together technical detail with biomedical relevance, this book offers an indispensable resource for students preparing to enter the field, for researchers seeking to expand their expertise, and for practitioners eager to understand the technologies shaping tomorrow's medicine.

In presenting both theoretical principles and practical case studies, the authors demonstrate not only mastery of the subject but also a vision for its future. That vision is one in which intelligent systems serve humanity responsibly, amplifying knowledge, improving lives, and addressing challenges that demand both scientific rigor and ethical care.

It is with great enthusiasm that I commend this book to its readers. May it serve as both a guide and an inspiration for the next generation of thinkers, innovators, and leaders in intelligent systems.

Via Giacomo Dina, Roma Dr. Leandro Pecchia

# Preface

This textbook provides a comprehensive introduction to intelligent systems, integrating foundational theories, learning paradigms, and practical applications with a specific emphasis on biomedical and healthcare domains. It is designed for students, researchers, and professionals in artificial intelligence, machine learning, and computational intelligence.

This book begins with an overview of intelligent systems, tracing the evolution of artificial intelligence (AI) and discussing various learning paradigms, including supervised, unsupervised, and reinforcement learning. It introduces the core principles of neural networks, comparing biological and artificial neurons, exploring different network architectures, and explaining key operations such as forward propagation and backpropagation.

Subsequent sections cover supervised learning techniques, including error-correction learning, support vector machines, decision trees, and gradient boosting. Reinforcement learning is introduced with Markov decision processes (MDP), Q-learning, and deep Q-networks. Unsupervised learning is examined in detail, covering Hebbian learning, clustering algorithms, autoencoders, and generative models.

This book then explores advanced neural network architectures, contrasting deep learning with traditional machine learning, investigating hybrid models, and detailing evaluation metrics for AI systems. A dedicated section focuses on fuzzy systems, covering fuzzy logic, fuzzy inference systems, and their integration with neural networks for expert decision-making and control applications.

The final theoretical section introduces modern AI and emerging technologies, including generative adversarial networks (GANs), transformers, AI in creative applications, and ethical considerations shaping the future of intelligent systems.

A dedicated section focuses on biomedical applications of intelligent systems, addressing AI-driven solutions in medical imaging, medical device performance assessment, clinical decision support, personalized medicine, remote health monitoring, robotic surgery, and wearable biosensors. Ethical and regulatory challenges specific to AI in biomedicine are also discussed.

By combining theoretical foundations with real-world applications, this textbook serves as an essential resource for understanding intelligent systems and their transformative role in modern AI-driven biomedical innovations.

Banja Luka, Bosnia and Herzegovina Dr. Almir Badnjević
Sarajevo, Bosnia and Herzegovina Lemana Spahić

# Acknowledgments

The authors wish to express their sincere appreciation to Sara Deumić, MSc, for her dedicated work in the process of formatting and technical finalization of this book. Her attention to detail and commitment to quality greatly contributed to the presentation and coherence of the final manuscript.

We are deeply grateful to our students, whose curiosity, enthusiasm, and insightful questions continuously inspire and motivate our academic work. Their engagement has been instrumental in shaping many of the ideas and examples presented throughout this book.

Finally, we extend our heartfelt thanks to our colleagues and families for their patience, understanding, and constant support during the demanding process of writing and editing. Their encouragement and belief in our work provided the foundation on which this project was completed.

# Contents

# List of Figures

# List of Tables

# Chapter 1
# Theories and Algorithms

This chapter provides a comprehensive foundation for intelligent systems, tracing their conceptual origins, computational models, and learning paradigms. It begins by defining intelligent systems and outlining their key characteristics, historical development, and core learning approaches—supervised, unsupervised, and reinforcement learning—along with hybrid methods, data representation, knowledge structures, and human–AI interaction principles (Sect. 1.1). Supervised learning methods are then introduced, covering regression, classification, and key algorithms such as linear regression, logistic regression, k-nearest neighbors, decision trees, and support vector machines, together with model evaluation and comparison (Sect. 1.2). Advanced supervised techniques follow, including ensemble learning, random forests, gradient boosting, hyperparameter tuning, strategies for imbalanced data, and transfer learning (Sect. 1.3). Unsupervised learning methods are presented next, focusing on clustering, dimensionality reduction, anomaly detection, and association rule discovery (Sect. 1.4), before moving to advanced approaches such as self-organizing maps, deep clustering, and manifold learning (Sect. 1.5). Chapter then explores the architecture and learning principles of perceptrons and multilayer perceptrons, highlighting representational power and training algorithms (Sect. 1.6), followed by the fundamentals of neural networks, including network types, loss functions, optimization, regularization, and convergence issues (Sect. 1.7). Specialized architectures are introduced with convolutional neural networks for image analysis and graph neural networks for structured data (Sect. 1.8). Fuzzy logic and fuzzy inference systems are examined as frameworks for reasoning with uncertainty and for integration with neural networks in control and decision support (Sect. 1.9). Chapter concludes with modern AI and emerging technologies, discussing generative AI, diffusion models, federated learning, explainable AI, and quantum machine learning as key drivers of future intelligent systems (Sect. 1.10).



A. Badnjević, L. Spahić, *Intelligent Systems in Biomedicine*, Series in BioEngineering, https://doi.org/10.1007/978-3-032-14785-1_1

## 1.1 Introduction to Intelligent Systems

The twenty-first century has witnessed a profound transformation in how technology interacts with human society. Intelligent systems, those capable of perceiving, reasoning, learning, and adapting, are no longer confined to research laboratories or speculative fiction. They permeate industries, healthcare, finance, education, and daily life. From digital assistants like Siri or Alexa to medical imaging software that aids in early cancer detection, intelligent systems constitute a defining technological paradigm of the modern era [1, 2].

At their core, intelligent systems aim to emulate aspects of human cognition. While early computers revolutionized computation by performing well-defined arithmetic and logical operations at scale, intelligent systems seek to go further: to deal with uncertainty, ambiguity, dynamic environments, and decision-making tasks that cannot be exhaustively pre-programmed. They represent a shift from **rule-based automation** toward **adaptive, learning-driven autonomy**. This chapter introduces the foundational principles, historical evolution, architectures, enabling technologies, and applications of intelligent systems [3, 4]. By the end, the reader will be equipped with a conceptual framework to understand not only how these systems work but also the societal implications they carry.

There is no single universally agreed-upon definition of an "intelligent system," but most converge around three capabilities:

1. **Perception**—acquiring data from the environment through sensors or inputs.
2. **Reasoning and Decision-making**—processing data to generate meaningful insights or actions.
3. **Learning and Adaptation**—improving performance over time without explicit reprogramming.

A working definition could be:

> ***An intelligent system is a computational entity capable of autonomously perceiving its environment, reasoning about data, learning from experience, and acting upon the environment to achieve defined goals, often in conditions of uncertainty.***

Core characteristics of intelligent systems are:

- **Autonomy**—independence from continuous human intervention.
- **Adaptivity**—ability to modify behavior in response to new circumstances.
- **Robustness**—tolerance of noise, errors, or incomplete information.
- **Goal-directed behavior**—alignment with objectives, whether externally specified or internally learned.

Traditional software executes predefined rules: if condition X is met, perform action Y. Intelligent systems, however, handle complex, dynamic, or unstructured environments where the rule space is too vast or unpredictable. For instance, autopilot software must continuously adapt to turbulence and unexpected changes rather than rely solely on rigid pre-coded responses.

Understanding intelligent systems requires tracing their origins through several key milestones in various disciplines (Fig. 1.1).

The idea of artificial intelligence predates computing. Philosophers like Aristotle pondered the formalization of reasoning. In the seventeenth century, René Descartes envisioned mechanistic models of thought. By the twentieth century, cognitive psychology and neuroscience contributed models of learning and perception that influenced system design.

Early AI systems focused on symbolic reasoning—manipulating abstract symbols through logical rules. Programs like **ELIZA** (a rudimentary natural language system) demonstrated potential but also revealed limitations, such as brittleness in real-world scenarios. As data availability grew, so did methods for learning from examples rather than explicit programming. Neural networks, genetic algorithms, and reinforcement learning emerged as frameworks for adaptive intelligence.

The last two decades have been defined by breakthroughs in **deep neural networks**, fueled by advances in computing power (e.g., GPUs), massive datasets, and algorithmic innovations [5, 6]. Applications such as speech recognition, image classification, and autonomous driving illustrate the maturity of intelligent systems today. Designing intelligent systems requires layered architectures that integrate multiple functional components (Fig. 1.2).

At the foundation of any intelligent system lies the **perception layer**, which is responsible for gathering and shaping raw input. In robotics, this may mean

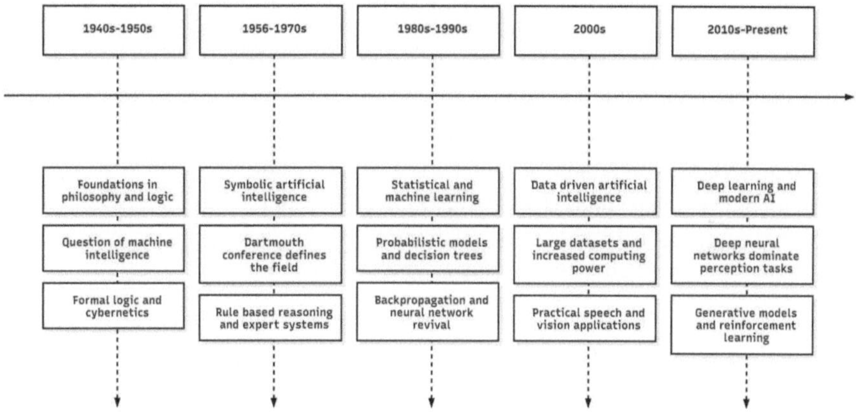


**Fig. 1.1** Timeline of intelligent systems

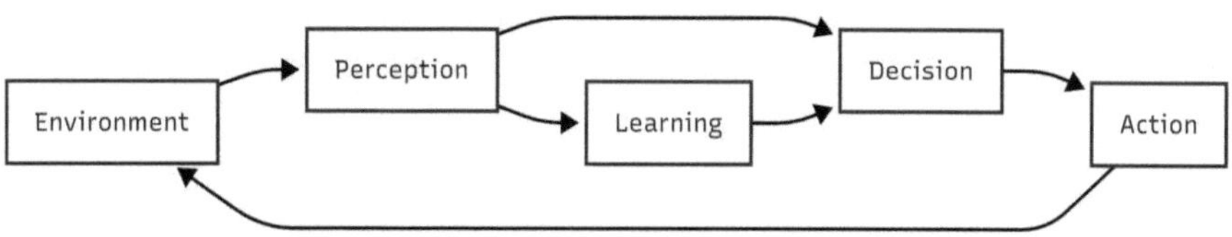


**Fig. 1.2** Conceptual diagram of an intelligent system architecture

cameras, LiDAR scanners, or tactile sensors that capture the surrounding environment. In software systems, it could be streams of data such as stock prices fluctuating in real time or the sequence of clicks a user makes while browsing a website. Before this information can be useful, it often undergoes preprocessing—steps like filtering noise, normalizing values, or extracting meaningful features from the raw input.

Building upon this is the **reasoning and decision layer**, where information is organized and made actionable. Knowledge may be structured through ontologies or semantic networks that define relationships between concepts, while uncertainty can be managed using probabilistic models such as Bayesian networks [7]. This layer is also where optimization and planning algorithms operate, charting courses of action and weighing alternatives based on the knowledge available.

The **learning layer** ensures that the system remains dynamic and evolves over time. Through supervised learning, it can handle labeled data and perform tasks such as medical diagnosis with increasing accuracy. Through unsupervised learning, it can uncover hidden patterns, such as grouping customers by purchasing behavior in marketing contexts. Reinforcement learning adds yet another dimension, enabling systems to improve through trial and error in sequential decision-making scenarios, as seen in game-playing agents or autonomous navigation.

Finally, insights must translate into tangible outcomes, which happens in the **action layer**. For physical systems, this may involve robotic actuators executing precise movements in the real world. For purely digital systems, it may manifest as automated decisions, approving a credit application, for instance, or adaptive interfaces that respond to a user's behavior in real time. In each case, the action layer closes the loop, turning perception, reasoning, and learning into practical impact.

Intelligent systems are ubiquitous:

- **Healthcare**: diagnostic imaging, drug discovery, robotic surgery.
- **Transportation**: autonomous vehicles, traffic management.
- **Finance**: fraud detection, algorithmic trading.
- **Education**: personalized learning platforms, adaptive testing.
- **Industry 4.0**: predictive maintenance, smart manufacturing (Fig. 1.3).

The capacity of an intelligent system to learn from data marks a profound departure from traditional rule-based programming. In the latter, every condition and response must be explicitly encoded by a human developer. This approach works adequately for well-defined domains, such as computing payroll or processing structured databases, but it collapses under the complexity of real-world environments [8]. A medical diagnostic system, for instance, cannot feasibly rely on an exhaustive set of rules written in advance; the combinatorial explosion of possible patient conditions, symptoms, comorbidities, and genetic variations is simply too vast. Machine learning emerged as a solution to this problem, enabling systems to infer patterns and make predictions directly from examples.

Machine learning can be understood as the science of constructing algorithms that improve their performance through experience. Instead of programming "if-then" statements, the designer provides the system with data, from which it extracts

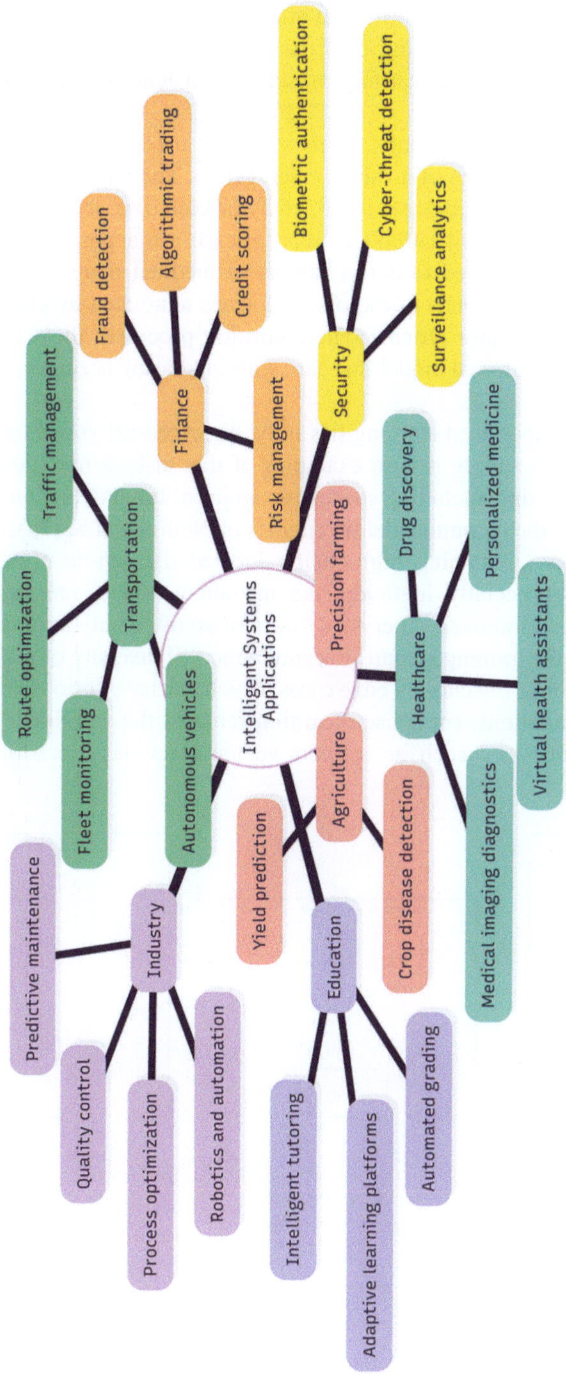


**Fig. 1.3** Application map of intelligent systems across domains

relationships that were not explicitly encoded. This ability to generalize beyond seen examples forms the heart of modern intelligent systems. While machine learning is a vast and rapidly evolving field, its paradigms can be grouped into three broad categories: supervised learning, unsupervised learning, and reinforcement learning.

**Supervised learning** is perhaps the most widely used paradigm and underpins many everyday applications of artificial intelligence. The term "supervised" refers to the fact that the learning process is guided by labeled data: inputs are paired with correct outputs, and the algorithm's task is to learn a mapping between the two. Consider a system designed to detect tumors in medical images [9, 10]. During training, it receives thousands of radiological scans annotated by experts as either "tumor present" or "tumor absent." The algorithm processes these examples and gradually refines an internal model that can, with increasing accuracy, classify new, unseen scans (Fig. 1.4).

The appeal of supervised learning lies in its directness: if you want a system to perform a task, you supply it with examples of that task. It has proven effective across various domains, including speech recognition, fraud detection, email spam filtering, and even the recommendation engines of Netflix and Spotify. Yet, it is not without limitations. Obtaining high-quality labeled datasets is often costly and time-consuming, especially in fields like medicine, where annotation requires expert knowledge. Moreover, supervised systems are often only as good as the data provided [11]. If the training set fails to capture the full diversity of real-world situations, the model may misclassify edge cases, which can have serious consequences. Despite these limitations, supervised learning remains the workhorse of machine intelligence. Techniques such as support vector machines, decision trees, and

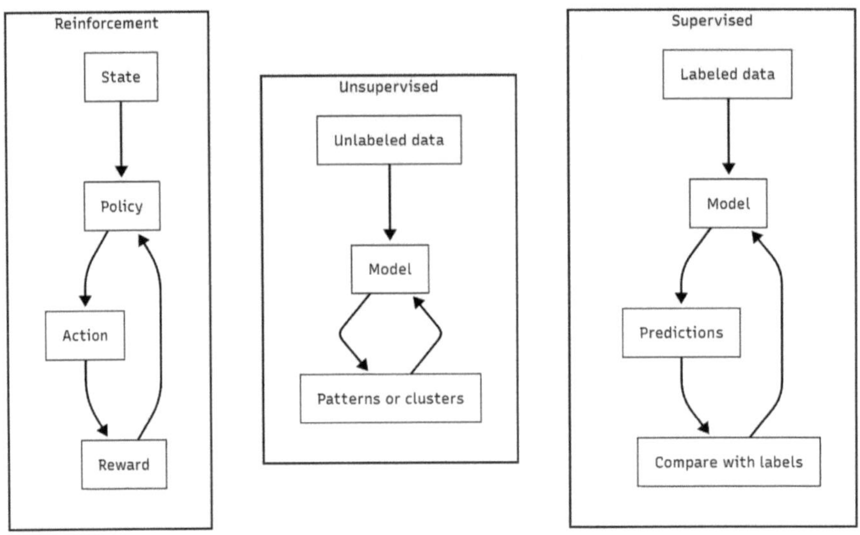


**Fig. 1.4** Diagram contrasting supervised, unsupervised, and reinforcement learning

especially deep neural networks have been refined within this paradigm. With the explosion of big data and cloud computing, supervised learning has achieved remarkable feats: large language models that power conversational agents, facial recognition systems used in security, and automated credit scoring in finance [12–15].

While supervised learning thrives on labeled data, the real world rarely provides such conveniently annotated examples. Vast amounts of information, from sensor readings to user behavior logs, arrive unlabeled, with no obvious structure. **Unsupervised learning** is designed to make sense of this raw data, uncovering hidden patterns, groupings, or structures without external guidance. A classic example is customer segmentation in marketing. Companies may have millions of purchase records, but no predefined categories describing customer types. By applying unsupervised algorithms such as clustering, the system can automatically identify groups of customers with similar buying behaviors [16]. These insights can inform targeted advertising campaigns, product recommendations, or strategic planning.

Unsupervised learning is not confined to business intelligence. In biology, it has been used to uncover subtypes of diseases based on genetic profiles, which can lead to more personalized medical treatments [17]. In cybersecurity, anomaly detection systems learn what constitutes "normal" network traffic and can then flag deviations that may indicate malicious activity. The power of unsupervised learning lies in discovery: it enables intelligent systems to find structure where none is obvious, revealing relationships that even human experts might overlook. However, evaluation can be tricky. Unlike supervised learning, where one can compare predictions to ground truth labels, unsupervised results often lack an objective benchmark [18]. The discovered clusters or patterns must be interpreted, and their utility judged in context.

**Reinforcement learning** represents a different approach altogether, inspired more by behavioral psychology than by traditional data analysis. In this paradigm, an agent learns not from static datasets but from interaction with an environment. The agent perceives the current state of the environment, takes an action, and receives feedback in the form of a reward or penalty. Over time, the agent aims to maximize cumulative reward, discovering strategies that yield the best long-term outcomes [19].

Perhaps the most famous demonstration of reinforcement learning is DeepMind's AlphaGo, the system that defeated world champion Lee Sedol in the game of Go, a feat once considered decades beyond reach. Unlike supervised learning, which might train on annotated game moves, reinforcement learning allowed AlphaGo to play millions of games against itself, gradually improving its strategy by trial and error. Similar principles drive robotics, where a machine may learn to walk, grasp objects, or navigate a room not by explicit programming but by repeated attempts guided by reward signals.

The paradigm's strength is its ability to tackle sequential decision-making problems, where actions have consequences that unfold over time. This makes it uniquely suited for domains such as robotics, resource management, and autonomous driving. Yet reinforcement learning is notoriously data-hungry [20]. Training an agent to mastery often requires millions of iterations, which may be impractical outside simulated environments. Moreover, designing the reward structure is delicate: poorly chosen rewards can lead to unintended behaviors, a phenomenon sometimes called "reward hacking."

The neat division into supervised, unsupervised, and reinforcement learning is pedagogically useful, but real-world intelligent systems increasingly blend these paradigms and introduce **hybrid approaches**. **Semi-supervised learning,** for example, leverages a small amount of labeled data alongside a large pool of unlabeled examples, striking a balance between practicality and accuracy. Self-supervised learning, a rapidly growing area, allows systems to generate their own labels from raw data, particularly in natural language processing. Here, models predict missing words in a sentence or the next frame in a video, effectively learning powerful representations without human annotation. Another promising direction is lifelong learning, where systems continuously update their knowledge across tasks rather than being trained once in isolation. This moves closer to the adaptability seen in human cognition. Similarly, meta-learning—often described as "learning to learn"—focuses on algorithms that can rapidly adapt to new tasks with minimal data, a capability vital for intelligent agents operating in dynamic, unpredictable environments.

Across all paradigms, one unifying theme is the central role of data and its representation. Raw data, whether pixels in an image or audio waveforms, is rarely suitable for direct use. Feature extraction, the transformation of data into a form amenable to learning. Remains a cornerstone of machine learning. In earlier decades, feature engineering was largely manual, requiring domain expertise to craft meaningful descriptors. The rise of deep learning, however, shifted this landscape, as neural networks learned hierarchical features directly from raw inputs. A convolutional neural network analyzing images, for instance, might automatically detect edges, shapes, and higher-order structures without explicit programming [21]. This automatic feature learning not only improved accuracy but also broadened the applicability of machine learning. Systems could now tackle tasks where feature engineering was previously infeasible. Yet, it also introduced new challenges, including the need for enormous datasets and computational resources [22]. Moreover, the resulting models, while effective, often operate as "black boxes," raising concerns about interpretability and trust.

The choice of learning paradigm is not merely a technical matter; it carries ethical implications. Supervised systems trained on biased datasets may perpetuate or even amplify societal inequities. Unsupervised clustering can inadvertently reinforce stereotypes if groups are defined along sensitive attributes. Reinforcement learning, when applied to areas like financial trading or online advertising, may optimize for profit at the expense of broader social good. Recognizing these risks, researchers increasingly emphasize fairness, accountability, and transparency in the design of machine learning systems (Fig. 1.5).

Moreover, the societal impact of machine learning is amplified by scale. A misclassified medical image might affect one patient, but a biased loan approval

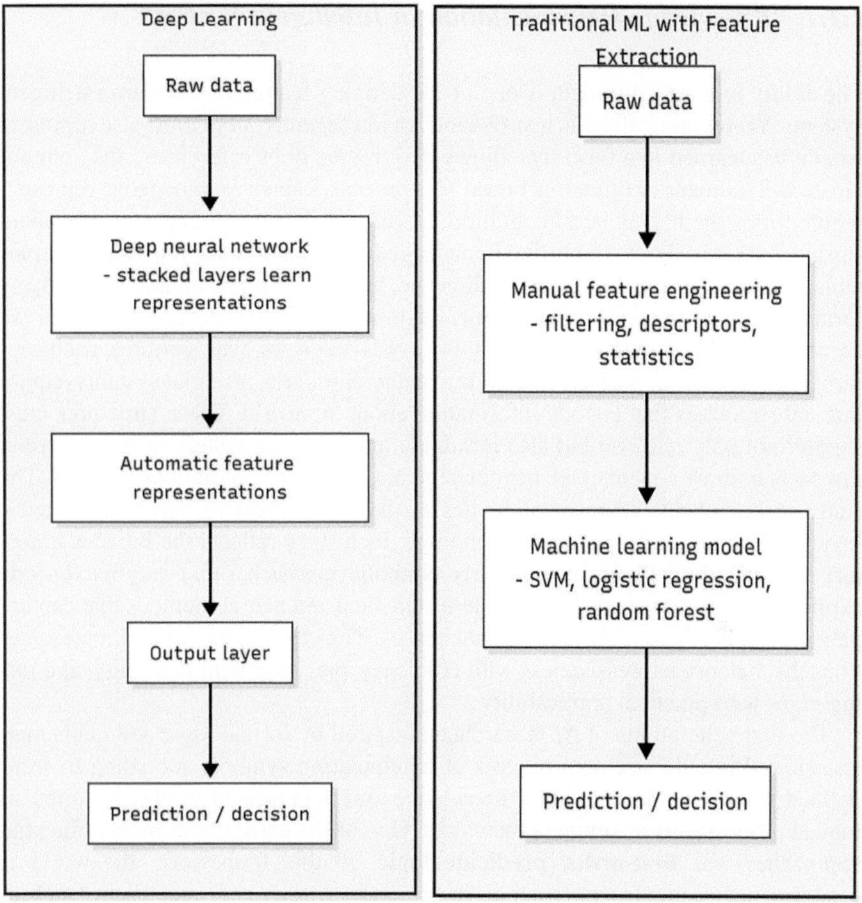


**Fig. 1.5** Feature extraction vs. deep learning

algorithm can influence thousands of lives. As intelligent systems become increasingly integrated into infrastructure, governance, and healthcare, the choice of paradigm, the quality of data, and the interpretability of models all assume critical importance. The landscape of machine learning paradigms continues to evolve. Advances in neuromorphic computing, hardware designed to mimic neural processes, promise more efficient learning. Bio-inspired algorithms, drawing from evolutionary principles or collective behaviors in nature, hint at new paradigms beyond the current triad [23]. At the same time, the integration of symbolic reasoning with data-driven learning seeks to combine the strengths of both approaches: the interpretability and structure of symbolic systems with the adaptability and scalability of machine learning.

### 1.1.1 Knowledge Representation in Intelligent Systems

The ability to learn from data is one of the defining features of modern intelligent systems. Yet learning alone is insufficient. An intelligent system must also represent what it has learned in a form that allows it to reason, draw inferences, and communicate with humans or other machines. This process, known as knowledge representation, forms the backbone of cognition in artificial systems. It is not merely about storing data; it is about structuring knowledge in ways that make it usable, interpretable, and transferable. In human intelligence, knowledge representation takes many forms: language, imagery, mathematical symbols, and embodied skills. When we describe a city to someone, we might use words, maps, or even gestures, each capturing different aspects of the underlying reality. Similarly, artificial systems require internal structures that encode information about the world. These structures must support not only retrieval but also reasoning, enabling the system to move beyond raw facts to draw conclusions, formulate plans, and make predictions [24, 25]. The study of knowledge representation lies at the intersection of computer science, logic, linguistics, and cognitive psychology. Its history reflects the broader trajectory of artificial intelligence: from early symbolic approaches that sought to encode explicit rules about the world, to modern statistical and neural methods that capture patterns implicitly in high-dimensional spaces. The challenge is to find representations that balance expressiveness with efficiency, precision with flexibility, and formal rigor with practical applicability.

The first generations of AI researchers, inspired by formal logic and mathematics, viewed intelligence as a process of manipulating symbols according to well-defined rules. In this paradigm, knowledge was represented explicitly, often as logical propositions or semantic networks. One of the earliest and most influential approaches was **first-order predicate logic**. In this framework, the world is described using predicates (such as *isHuman(Socrates)*) and quantifiers (such as "for all x, if x is human, then x is mortal"). Reasoning then proceeds through formal inference rules, such as modus ponens, which allow the system to derive new facts from existing ones [26]. This symbolic paradigm had several strengths. It offered clarity, interpretability, and a firm foundation in mathematical logic. Systems could provide explicit justifications for their conclusions, making them transparent in ways that many modern machine learning models are not. Early expert systems, such as MYCIN for medical diagnosis, relied heavily on symbolic knowledge representation, encoding thousands of rules provided by domain experts. Yet symbolic approaches also revealed serious limitations. The real world is messy, uncertain, and often resistant to clean logical formalization [27]. Knowledge expressed as rules can quickly become brittle, failing when faced with exceptions or incomplete information. For example, while a rule might state "birds can fly," it requires numerous exceptions to account for penguins, ostriches, or injured animals. This "frame problem" illustrated how difficult it was to capture commonsense knowledge in purely symbolic terms (Fig. 1.6).

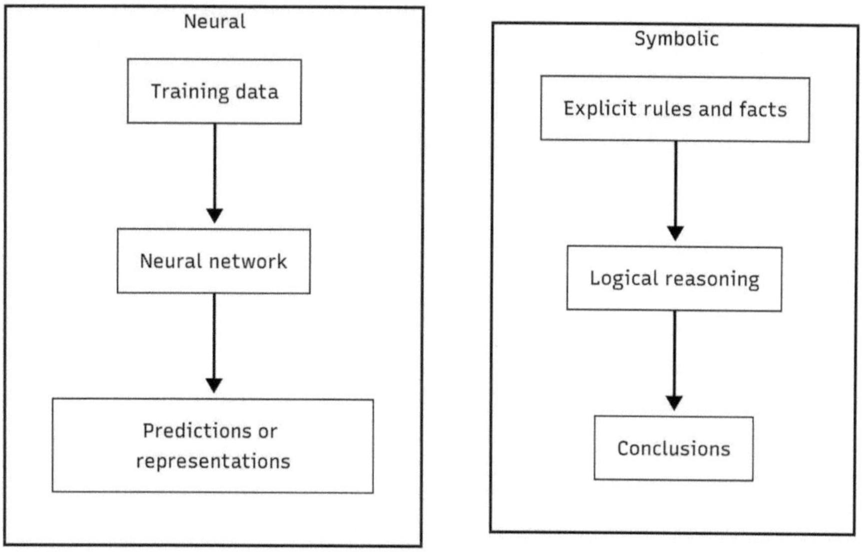


**Fig. 1.6** Comparison of symbolic vs. neural knowledge representations

In response to these challenges, researchers developed more flexible forms of representation. **Semantic networks** represented knowledge as nodes (concepts) connected by edges (relationships). For instance, a network might connect the concept "dog" to "mammal" via an "is-a" relationship, and to "barks" via a "has-behavior" link. These networks captured hierarchical relationships and allowed inheritance: if a dog is a mammal, and mammals are warm-blooded, then dogs inherit the property of being warm-blooded.

Building on this idea, Marvin Minsky introduced the concept of **frames**, structured collections of attributes that describe stereotypical situations or objects. A "restaurant" frame, for example, might include slots for "waiter," "menu," "meal," and "bill." By filling in these slots with specific values, the system could reason about particular instances of the general concept. Frames provided a more natural way to encode commonsense knowledge, as they mirrored the human tendency to think in terms of prototypes and schemas. Despite their intuitiveness, semantic networks and frames still struggled with ambiguity and exceptions. They also lacked the rigorous inferential machinery of formal logic, limiting their ability to support complex reasoning [28]. Nevertheless, they represented a significant step toward more nuanced and human-like knowledge representation. The rigidity of symbolic methods led to the incorporation of uncertainty into knowledge representation. Real-world environments rarely present information in crisp, binary terms; instead, they are characterized by noise, ambiguity, and partial evidence. To address this, researchers turned to probabilistic and fuzzy approaches. **Bayesian networks** emerged as a powerful tool in the 1980s and 1990s. These are graphical models that represent variables as nodes and probabilistic dependencies as edges. For example, a Bayesian network for medical diagnosis might link “smoking” to “lung cancer” and “cough,” with probabilities quantifying the strength of these associations. When new evidence is observed, for instance, a patient develops a persistent cough, the network can update its beliefs about the likelihood of lung cancer using Bayes' theorem. Similarly, **fuzzy logic** introduced degrees of truth, moving beyond the binary true/false dichotomy of classical logic. A temperature of 28 °C might be “somewhat hot” rather than simply “hot” or “not hot.” Fuzzy systems proved effective in control applications, such as washing machines or air conditioning systems, where rules based on vague concepts like “a little more” or “quite large” reflect human reasoning. Probabilistic and fuzzy representations acknowledged the uncertainty inherent in real-world reasoning, providing intelligent systems with the ability to make decisions under incomplete or noisy conditions. However, they often required carefully crafted models and could become computationally expensive as the number of variables increased.

With the rise of machine learning, knowledge representation underwent another transformation. Rather than being explicitly encoded by human designers, knowledge began to be captured implicitly in the parameters of statistical models or neural networks. In a neural network trained to recognize images, for example, knowledge about visual categories such as “cat” or “dog” is distributed across millions of numerical weights (Fig. 1.7).

These weights encode features ranging from simple edges to complex object shapes. Similarly, in large language models, knowledge about grammar, semantics, and even world facts emerge from patterns in the parameters learned from vast corpora of text. This shift brought enormous practical success. Neural representations proved highly effective for tasks where explicit symbolic models were intractable, such as natural language understanding or real-time vision. They could capture subtleties and nuances that rule-based systems struggled with, and they scaled well with data. Yet this power came at a cost: interpretability. Unlike symbolic systems, which could explain their reasoning in human-readable rules, neural networks often function as opaque “black boxes.” Their knowledge is embedded in high-dimensional weight matrices that defy straightforward interpretation. This opacity has sparked

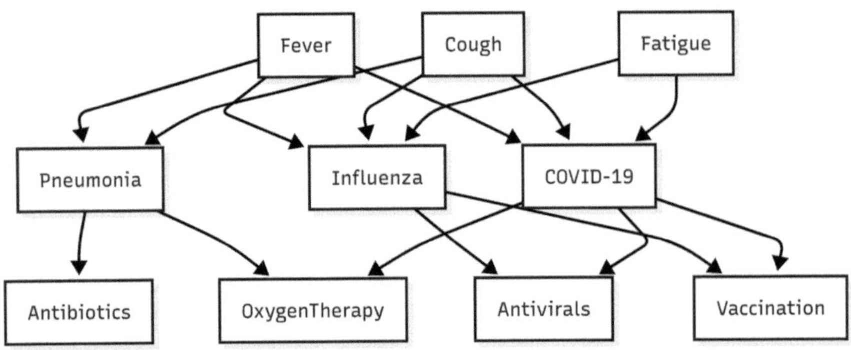


**Fig. 1.7** Simplified knowledge graph

intense debate in fields such as medicine and law, where transparency and accountability are crucial.

Recognizing the strengths and weaknesses of both symbolic and statistical approaches, researchers have increasingly pursued hybrid models. These systems aim to combine the explicit reasoning capabilities of symbolic representations with the adaptive learning of neural networks [29, 30]. One promising avenue is **neuro-symbolic AI**, where neural networks handle perception and pattern recognition, while symbolic components provide logical reasoning and structure. For instance, a neuro-symbolic system for visual question answering might use a neural network to identify objects in an image, and then apply symbolic reasoning to answer a query such as "Is there a red ball next to the dog?" Hybrid representations also hold promise for achieving better interpretability. By linking neural outputs to symbolic structures, they can provide human-readable explanations while retaining the flexibility of statistical learning. Such approaches may be crucial for bridging the gap between high-performing but opaque models and the need for transparent, trustworthy intelligent systems.

In recent years, **ontologies** and **knowledge graphs** have emerged as central tools for structuring knowledge in intelligent systems. An ontology provides a formal specification of concepts and their relationships within a domain, often expressed in languages such as OWL (Web Ontology Language). Knowledge graphs extend this idea, representing entities and their interconnections as large-scale graphs. The most prominent example is the Google Knowledge Graph, which underpins search by connecting entities like "Barack Obama," "President," and "United States" through explicit relationships. When a user searches for "Obama," the system not only retrieves documents containing the word but also understands the entity and its attributes. Knowledge graphs support semantic reasoning, linking structured and unstructured data across vast domains. They are used in healthcare to connect symptoms, diseases, drugs, and treatments; in finance to map companies, transactions, and risks; and in e-commerce to power recommendation engines [31]. By integrating ontologies with machine learning, systems can both reason over explicit knowledge and adapt to new data, achieving a richer form of intelligence.

The way knowledge is represented in intelligent systems has profound implications for society. Symbolic systems risk reflecting the biases of their designers, embedding subjective rules into seemingly objective frameworks. Statistical systems risk reproducing biases in data, reinforcing inequities at scale. Knowledge graphs raise questions of authority and ownership: who decides which entities and relationships are encoded, and whose knowledge is excluded?

At a deeper level, knowledge representation confronts philosophical questions about the nature of knowledge itself. Can intelligence be reduced to symbols and probabilities, or does it require embodiment and context? To what extent do artificial representations approximate human understanding, and where do they diverge? These questions remind us that knowledge representation is not a solved technical problem but an ongoing negotiation between computational formalism, human cognition, and social values. The future of knowledge representation likely lies in continued integration. Symbolic, probabilistic, and neural methods each offer unique strengths: explicit reasoning, uncertainty handling, and adaptive learning. Their combination, mediated through ontologies, knowledge graphs, and neuro-symbolic architectures, may yield systems capable of both high performance and human-like interpretability [32]. Moreover, advances in cognitive neuroscience and philosophy may inspire new paradigms, moving beyond the symbolist versus connectionist divide. Embodied and enactive theories of cognition suggest that knowledge is not merely stored but enacted through interaction with the world. Intelligent systems of the future may therefore represent knowledge not only internally but also through their actions, behaviors, and participation in social contexts. Ultimately, knowledge representation is the fabric upon which intelligence is woven. It is what allows raw data to become structured information, information to become actionable insight, and insight to be communicated and shared. For intelligent systems to be trustworthy, adaptable, and aligned with human values, their representations of knowledge must be as rich, nuanced, and ethically grounded as the societies they are designed to serve.

### *1.1.2 Human–AI Interaction and Interpretability*

Intelligent systems do not exist in isolation. They are designed, deployed, and evaluated within human contexts, whether assisting a doctor in making a diagnosis, helping a student learn a new subject, or recommending financial investments. The effectiveness of such systems is therefore not measured solely by computational performance but also by how well they interact with people [33]. **Human–AI interaction** has become a critical dimension of intelligent system design, encompassing issues of usability, trust, transparency, and interpretability.

The history of computing provides a useful perspective. Early mainframe computers were operated by specialists who interacted with them through punch cards and command-line instructions. With the advent of graphical user interfaces, personal computing became accessible to non-experts, fundamentally changing the role of computers in society. A similar transformation is underway with intelligent systems: they must move from being opaque tools in the hands of technical experts to becoming accessible, trustworthy collaborators for a broad range of users. Perhaps the most pressing issue in human–AI interaction is interpretability—the extent to which a system's operations and decisions can be understood by humans. Interpretability is not merely an aesthetic preference; it is a prerequisite for trust. A physician will be reluctant to adopt a diagnostic tool if it cannot explain why it flagged a particular image as cancerous [34]. A judge will hesitate to use an algorithmic risk assessment system if it cannot justify its recommendations. Without interpretability, human users are left in a position of blind faith, which is antithetical to responsible decision-making in critical domains. Interpretability can take different forms. In some cases, it involves **transparency**—exposing the inner workings of the model itself. Decision trees, for instance, are inherently interpretable because they express reasoning as a series of if–then rules. In other cases, interpretability is achieved through **post hoc explanations**—techniques that approximate the reasoning of complex models without fully exposing their internal mechanisms. Methods such as Local Interpretable Model-agnostic Explanations (LIME) or SHapley Additive exPlanations (SHAP) provide insights into which features influenced a prediction, even if the underlying model is a neural network with millions of parameters. Yet interpretability is not a purely technical problem. Different stakeholders require different kinds of explanations. A data scientist may want detailed information about model weights and feature interactions, while a clinician may prefer a concise rationale expressed in plain language. Designing explanations that are accurate, comprehensible, and context-appropriate is a central challenge for human–AI interaction (Fig. 1.8).

Trust in intelligent systems is shaped not only by interpretability but also by psychological and social factors. Research in human–computer interaction has shown that people often anthropomorphize machines, attributing intentions or emotions to them even when these do not exist. A conversational agent that uses natural language may be perceived as more trustworthy or competent simply because it mimics human communication. At the same time, humans are prone to cognitive biases that can distort their evaluation of AI systems. **Automation bias** refers to the tendency to over-rely on automated systems, accepting their outputs even in the face of contradictory evidence. In aviation, pilots have sometimes failed to override faulty autopilot systems, leading to accidents. Conversely, **algorithm aversion** describes the opposite tendency: people may distrust automated systems after observing even minor errors, while tolerating similar errors from human counterparts. Both extremes, over-reliance and under-reliance, can undermine the effective integration of intelligent systems into human workflows [35, 36]. Designing for calibrated trust is therefore essential. Users must neither treat AI outputs as infallible nor dismiss them outright. Achieving this balance requires not only technical accuracy but also careful interface design, transparent communication of uncertainty, and ongoing user education. Rather than viewing intelligent systems as replacements for human decision-makers, a growing body of research emphasizes their role as collaborators. In this model, the strengths of humans and machines are seen as complementary. Machines excel at processing vast quantities of data, identifying subtle statistical patterns, and executing repetitive tasks with precision and accuracy. Humans, by contrast, bring contextual understanding, ethical judgment, and the ability to reason across domains. In medicine, this complementarity is evident in diagnostic support systems. A radiologist, aided by an AI that highlights suspicious regions in a scan, can combine the system's pattern recognition with their own clinical knowledge and intuition. In law, algorithms may surface relevant

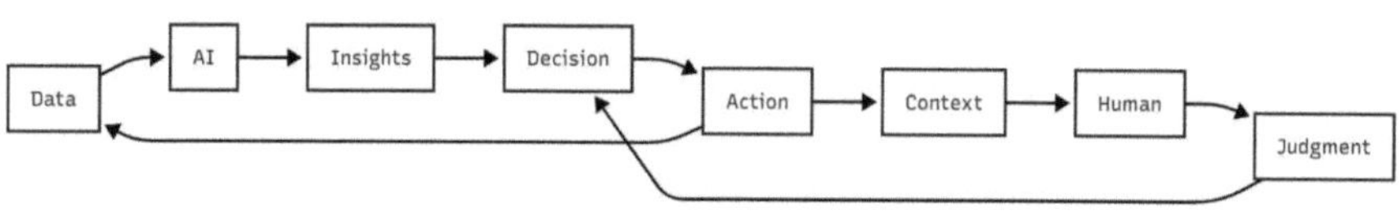


**Fig. 1.8** Schematic of human–AI collaboration

precedents, but the interpretation and application of these precedents to novel cases remain human responsibilities. The design challenge is to create interfaces and workflows that amplify these synergies rather than creating friction. This vision aligns with the concept of **human-centered AI**, which prioritizes human values, needs, and agency in system design. Human-centered AI resists the narrative of replacement and instead frames intelligent systems as augmentations of human capability. This requires not only technical adjustments but also cultural shifts in how organizations integrate AI into their operations (Fig. 1.9).

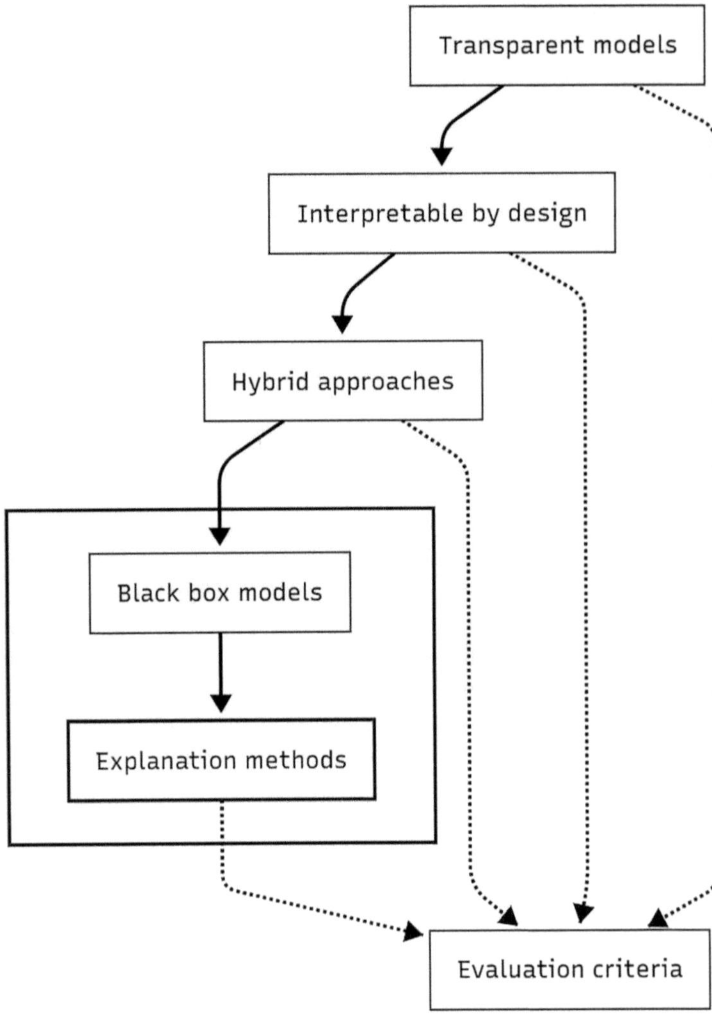


**Fig. 1.9** Diagram illustrating interpretability spectrum

A distinctive challenge in human–AI interaction is the communication of uncertainty. Intelligent systems rarely operate with absolute certainty; their outputs are typically probabilistic estimates. A medical diagnostic system might estimate an 85% probability of pneumonia, or a self-driving car might assess a 70% chance that an object ahead is a pedestrian. How should such uncertainty be communicated to users? Presenting probabilities directly may overwhelm or confuse non-experts, who often misinterpret statistical information. Simplifying uncertainty into categorical outputs ("disease present" vs. "disease absent") risks giving a false impression of certainty. Research suggests that effective communication requires tailoring to the user's expertise and decision-making context. For some users, visualizations such as confidence intervals or heatmaps may be helpful; for others, qualitative statements ("the evidence moderately supports this conclusion") may be more appropriate. Failure to communicate uncertainty effectively can have serious consequences. Overstating confidence may lead to unwarranted reliance on the system, while understating it may cause users to ignore valuable insights. Developing standards for uncertainty communication remains an open area of research, particularly in safety-critical domains.

Human–AI interaction is not limited to usability and interpretability; it also carries ethical dimensions. The ways in which systems present information can influence decisions, raising concerns about manipulation, bias, and the fairness of these decisions. For example, recommendation algorithms on social media platforms shape the information users encounter, often reinforcing echo chambers and polarizing discourse. Transparency alone does not resolve these issues. A biased system that explains its reasoning remains biased [37]. Addressing ethical concerns requires a systemic approach: auditing datasets, monitoring outcomes, and involving diverse stakeholders in the design of the system. Moreover, human–AI interaction must account for power dynamics. If a system presents recommendations in ways that pressure users into compliance, the human's role as the ultimate decision-maker is undermined. Ensuring meaningful human control requires both technical safeguards and institutional oversight.

A recurring tension in the design of intelligent systems is the trade-off between interpretability and performance. Simpler models, such as linear regressions or decision trees, are easy to interpret but may lack the accuracy of deep neural networks. Conversely, highly complex models often achieve state-of-the-art performance but resist straightforward explanation. Researchers have proposed several strategies to navigate this trade-off. One is to develop inherently interpretable models that aim to match the performance of black-box models while retaining transparency. Another approach is to enhance post hoc explanation methods, making complex models more comprehensible without altering their internal structure. A third approach is to combine interpretable models for high-stakes decisions with complex models for low-risk applications.

The appropriate balance depends on the domain. In advertising, where errors may have minor consequences, performance may take precedence over interpretability. In healthcare or criminal justice, interpretability is paramount, even at some cost to predictive accuracy. Recognizing these contextual differences is essential for responsible deployment. As intelligent systems become more integrated into daily life, their modes of interaction extend beyond text and graphics. Voice assistants, such as Alexa or Siri, allow for natural language interaction, while gesture recognition and haptic feedback enable more embodied forms of communication. In robotics, interaction becomes fully physical, as machines share space with humans and must navigate issues of safety, coordination, and social cues. Multimodal interaction, combining speech, vision, touch, and other modalities, holds promise for more natural and intuitive interfaces. A user might ask a robot to "pick up that red cup," while pointing to the object; the system must integrate speech and gesture to interpret the command. Such interactions move closer to human–human communication, but they also introduce new challenges, including disambiguation, context awareness, and cultural variation.

Looking forward, the trajectory of human–AI interaction suggests several key developments. First, interpretability will become increasingly central as intelligent systems are deployed in high-stakes domains. Regulatory frameworks may mandate explainability, much as financial systems are required to maintain audit trails. Second, collaboration models will continue to evolve, with intelligent systems acting as teammates rather than tools. This raises questions of responsibility: if a human and AI jointly make a decision, who is accountable for the outcome? Third, personalization will become a defining feature. Just as modern websites adapt interfaces to individual users, future intelligent systems may tailor their explanations, uncertainty communication, and interaction styles to each user's preferences and expertise. This personalization promises greater effectiveness but also raises concerns about manipulation and loss of autonomy. Finally, advances in neuroscience and cognitive science may inspire new paradigms of interaction, emphasizing embodied and social aspects of intelligence. Intelligent systems may not only respond to human inputs but also anticipate needs, detect emotional states, and engage in collaborative problem-solving [38]. The line between tool and partner may blur further, demanding careful ethical reflection.

Human–AI interaction and interpretability are not ancillary concerns but central dimensions of intelligent systems. They determine whether technical advances translate into real-world impact, whether users adopt and trust new tools, and whether society can hold systems accountable for their actions. The challenge is to design systems that are not only powerful but also understandable, trustworthy, and aligned with human values. As intelligent systems continue to permeate medicine, law, education, and daily life, their success will depend less on raw computational prowess and more on their ability to interact meaningfully with people. Interpretability, trust, collaboration, and ethical design are therefore not optional features but foundational principles of the next generation of intelligent systems.

## 1.2 Supervised Learning Techniques

A supervised learning method is a machine learning approach in which artificial intelligence models learn from pre-existing, human-labeled input and output datasets. When provided with new, unlabeled real-world input data, the trained model gets it right because it has learned the underlying input and output relationships. Example data points and the appropriate outputs or responses make up labeled data [39, 40]. The algorithm changes its weights as input data are provided one by one until the model gets properly fitted. The model is explicitly trained to recognize patterns between features and data labels by being trained with labeled training data. Unsupervised learning, as opposed to supervised learning, employs data without labeled outputs [41]. Recognizing patterns or groupings in data—such as grouping similar data points together—would be the general purpose of unsupervised learning (Fig. 1.10).

We have two general categories of Supervised Learning in Machine Learning

1. Regression: Returns a continuous value (price, temperature)
2. Classification: Returns a class label (spam/not spam)

If the output variable is a class or category, then classification. It is possible to train an email filtering system, for instance, to classify incoming mail as "spam" or "not spam" based on features like sender address or subject line. In the case of the output being a numeric value, regression is applied. Predicting home prices based on input attributes like location, bedrooms, and square footage is one well-known example [42]. Building a model that can effectively generalize to new, unseen data after learning from labeled training data is the goal in both instances.

### 1.2.1 Linear Regression

**Linear regression** is a statistical method of finding a straight line that explains a mathematical relationship between the independent variable and the dependent variable. For example, if we are attempting to predict a student's performance in an exam based on the number of hours they have studied. We observe that the greater the number of hours students study, the better their grades improve [43, 44]. Therefore, the independent variable (input) would be the number of hours studied, because it is the factor we can monitor or control. The dependent variable(output) would be the test score, as it varies with the number of hours studied (Fig. 1.11).

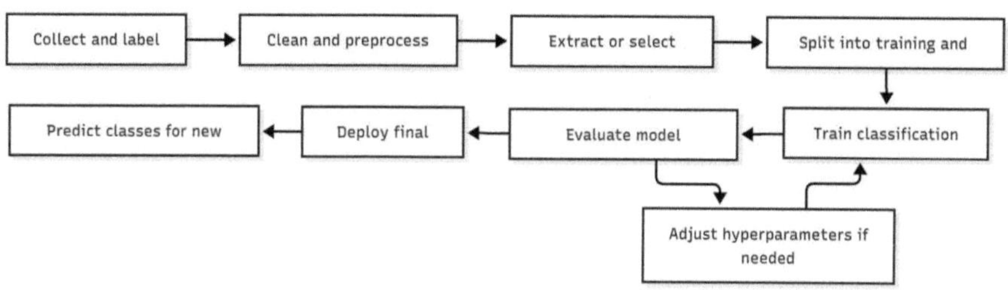


**Fig. 1.10** Supervised learning classification flowchart

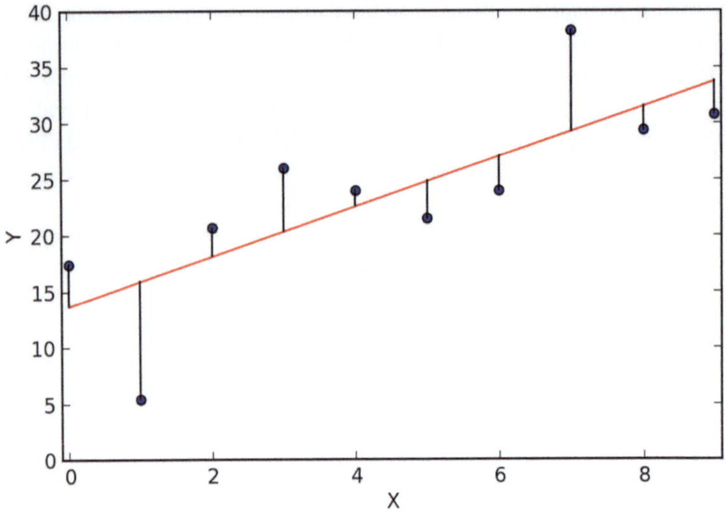


**Fig. 1.11** Best fit line of a linear regression model (https://commons.wikimedia.org/wiki/File:Residuals_for_Linear_Regression_Fit.png)

For simple linear regression (one independent variable), the **best-fit line** can be expressed by the equation:

$$y = mx + b \tag{1.1}$$

where:

- y is the predicted value (dependent variable)
- x is the input (independent variable)
- m is the gradient of the line (what varies as x varies)
- b is the intercept (y when x = 0)

The model learns the best values of m and b from the training data using a method known as least squares, which determines the values of m and b that minimize the sum of squared differences between the actual and predicted values [45]. The differences are called residuals. The formula for residuals is:

$$r_i = y_i - \hat{y}_i \tag{1.2}$$

where:

- $y_i$ is the observed value
- $\hat{y}_i$ is the line predicted value for that $x_i$

The model tries to minimize prediction error with a loss function:

$$MSE = \frac{1}{n}\sum_{i=1}^{n}\left(y_i - \hat{y}_i\right)^2 \quad (1.3)$$

where:

- $y_i$ is the real value
- $\hat{y}_i$ is the predicted value
- n is the number of points (Table 1.1)

**Disadvantages and Limitations**

**Limitations**

- Restaurant with good feature engineering
- Supposes a linear relationship between variables
- Tends to underfit or overfit data

**Advantages**

- Not difficult to understand and easy
- Fast to train even on large databases
- Run as a baseline model the majority of the time [46]

## *1.2.2 Logistic Regression*

**Logistic regression** is applied whenever the output variable is categorical or if it falls in one of a small set of classes. For example, based on the number of study hours, it can be used to determine whether a student will pass or fail an exam [47]. Logistic regression places the output in either two (or more) possible outcomes after the probabilities have been estimated, while linear regression estimates quantities.

**Table 1.1** Summary of Linear Regression

| Aspect | Description |
|---|---|
| Purpose | Predict a continuous outcome based on one or more input variables |
| Type of model | Parametric, supervised learning |
| Key assumption | Linear relationship between predictors and target |
| Equation | $y = \beta_0 + \beta_1 x1 + \cdots + \beta_n x_n + \varepsilon y$ |
| Parameters | Intercept ($\beta_0$), coefficients ($\beta_i$) |
| Training method | Ordinary least squares (minimizes sum of squared errors) |
| Evaluation metrics | R-squared, adjusted R-squared, mean squared error (MSE), root mean squared error (RMSE) |
| Strengths | Simple, interpretable, efficient for small to medium datasets |
| Limitations | Sensitive to outliers and multicollinearity; assumes linearity and homoscedasticity |

There are three types of logistic regression:

1. Binomial: two and only two kinds of dependent variables are feasible, i.e., 0 or 1, pass or fail
2. Multinomial: dependent variable can be one of three or more unordered categories, i.e., 'cat', 'dogs', 'sheep'.
3. Ordinal: greater than three ordered dependent variable types, i.e., 'low', 'medium', or 'high', are feasible.

The core function of logistic regression is the **Sigmoid Function**. It transforms any real value ranging from 0 to 1 to another value. The logistic regression curve is like an "S" since the value must be between 0 to 1, which is the maximum value [48].

The sigmoid function or logistic function is also known as the S-form curve (Fig. 1.12).

### *1.2.3 K-Nearest Neighbors (k-NN)*

**K-Nearest Neighbors (KNN)** is a supervised learning technique that can be used for classification and regression problems. After finding the "k" nearest data points for a given input, it makes a prediction based on the class with the highest count if it is a classification problem or based on the average value if it is a regression problem. Since KNN does not assume anything about the distribution of data itself, it is an instance-based, non-parametric learning algorithm [49]. K-Nearest Neighbors is what can be termed a lazy learner algorithm because it keeps the data set and then performs some operation on it in the classification process later, instead of learning from training data directly.

Applications of k-NN in machine learning

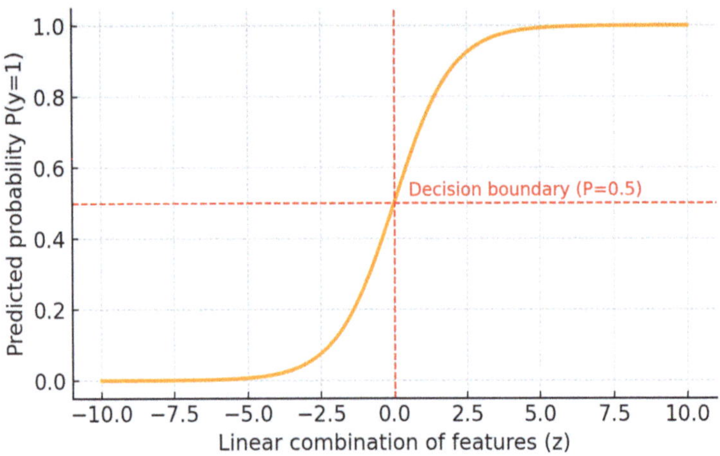


**Fig. 1.12** Sigmoid curve

- Processing data: Imputation of missing values is the process by which the k-NN algorithm predicts the missing values that are normally present in datasets.
- Recommendation Systems: The k-NN algorithm has been used to automatically suggest extra content to users based on clickstream data on websites.
- Finance: Many financial and economic applications have also used it. Banks, for example, are able to estimate the risk of lending to an individual or a company by applying k-NN to credit data. It is used to evaluate the creditworthiness of a loan applicant.
- Healthcare: k-NN has also been used in the healthcare sector for predicting the risk of prostate cancer and heart attacks. The algorithm performs the most likely gene expressions.
- Pattern recognition: k-NN has also been used to recognize patterns, for example, in text and number classification. This has proved useful to recognize handwritten numbers that you see on envelope addresses or forms.

The 'k' in the k-NN refers to the number of neighbours or closest points that the algorithm will use while deciding. For instance, if a fruit is known by its shape and size. It would be compared with the known fruit [50, 51]. Therefore, the algorithm would consider the three closest fruits to the new one if k = 3. The algorithm determines the new fruit to be an apple when two of them are apples and one is a banana, because apples are the majority among their neighbors. k-NN employs distance measures to determine the nearest neighbor. Such neighbors are used to perform regression and classification. We employ the distance measures:

- The Euclidean Distance: Euclidean distance is the straight line distance between two points in a space or plane. It is similar to the shortest distance that is traveled by going in a straight line from one place to another.
- Manhattan distance: if horizontally and vertically, you could move in straight lines, as in a grid or city street, the distance would all be covered. Also called 'taxicab distance', as taxis can travel only the grid-like roads of the city in New York.
- Minkowski distance: Euclidean and Manhattan distances are both special cases of the Minkowski distance, which is a member of a family of distances. It is a general formula that, for various values of p, can be made to represent both the Manhattan distance and the Euclidean distance.

**Advantages**

- Easy to use: New data scientists will learn this algorithm as one of the first neural networks they encounter due to its accuracy and simplicity.
- Highly adaptable: since all training data are kept in memory, the algorithm picks up new data as new data are added.
- Fewest number of hyperparameters: in contrast to other machine learning algorithms, k-NN has only a distance metric and k.

**Drawbacks**

- Unstable: k-NN is more resource-intensive when it comes to data storage and memory in comparison to other classifiers because it is a lazy algorithm. From an economic and timing perspective, this can be expensive. Extra storage and memory will increase the cost of operations, and processing large volumes of data takes longer

- Curse of dimensionality: Because of the curse of dimensionality, the k-NN algorithm generally does not work well when dealing with high-dimensional data inputs. The peaking phenomenon is also referred to as this phenomenon that occurs when the algorithm attains the best number of features and then adds more features, which brings the number of classification errors, especially when there is a small sample size.
- Most likely to overfit: k-NN is also most likely to overfit because of the 'dimensionality curse'.

While **dimensionality reduction and feature selection** methods are used to stop this from happening, the value of k can affect the performance of the model as well [52]. Although higher k values are likely to 'smooth out' the prediction values since they are averaging over a wider area, or 'neighborhood,' lower values for k are apt to overfit the data. Similarly, it will underfit the data if k is excessively large.

### *1.2.4 Decision Trees*

**Decision trees** are among the most well-liked methods in supervised learning due to their understandable structure, interpretability, and ease of use with various kinds of data. Their roots trace back to statistical classification research in the 1960s and 1970s; however, they were primarily developed during the 1980s, when important algorithms such as **ID3 (Iterative Dichotomiser 3)**, **CART (Classification and Regression Trees)**, and subsequently **C4.5** and **C5.0** were created.

Essentially, decision trees aim to replicate human decision-making by organizing decisions in a hierarchical, sequential manner [53, 54]. They learn from training data that is labeled by creating a series of decision rules that split the data into progressively more homogeneous subsets. In supervised learning, this equates to inducing a mapping from input features X $(x_1, x_2, \ldots, x_n)$ to a target variable Y, which can be in the form of discrete classes (classification) or continuous values (regression).

In classification tasks, the decision space is split into regions corresponding to a dominant class, typically based on measures such as **information gain** or **Gini impurity**. In regression, the tree attempts to minimize variance in each leaf node by predicting mean target values for subsets. Such transparency in decision logic makes decision trees ideal for use in applications that require interpretable model outputs, such as clinical diagnosis, financial forecasting, and regulatory decision systems.

Moreover, decision trees are the computational engine for ensemble methods like **Random Forests** and **Gradient Boosting Machines,** which improve generalization and robustness through the combination of a large number of base trees' predictions. These state-of-the-art techniques retain the interpretability advantages of decision trees while overcoming the drawbacks of single trees, such as overfitting and sensitivity to data variation (Fig. 1.13).

A decision tree represents decision-making processes by a hierarchical, tree-like model that reflects different decisions and their possible outcomes. The basic components of a decision tree are:

- Root node: It is the topmost node, which is the most significant feature chosen for the split of the dataset. The selection of this feature is performed using measures such as information gain maximization or impurity minimization.
- Internal nodes: These are the intermediate decision-making points where the dataset is further separated based on some feature conditions.
- Branches (edges): Branches are the possible outcomes or decisions from a node. Each branch connects one node to another, either to other internal nodes or to terminal nodes, and is labeled with a specific value or condition of the considered feature.
- Leaf nodes (terminal nodes): They are the final outcomes of the decision tree. They provide class labels for classification problems and numerical predictions, typically the mean of target values in the subset, for regression problems.

Each path from the root node to a leaf node can be interpreted as a decision rule consisting of a conjunction of conditional statements on feature values. For instance, a decision tree predicting whether or not a student passes a course might begin by asking if "hours studied > 4." Depending on the result, the tree might then ask "attendance > 75%," and so on until it makes a final prediction (e.g., "Pass" or "Fail"). The tree is built on the basis of a recursive partitioning algorithm that successively splits the data into more homogeneous subsets in relation to the target

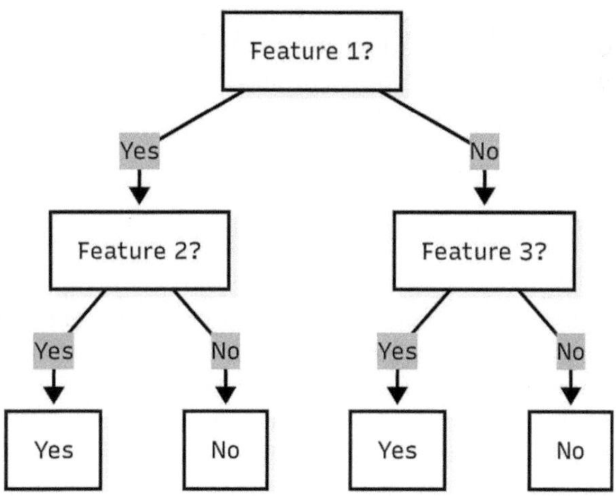


**Fig. 1.13** Decision Tree Structure (https://commons.wikimedia.org/wiki/File:Decision_Tree_Depth_2.png)

variable. The recursive splitting enables the model to discover complex decision boundaries in the data.

One of the main tasks in the decision tree building process is the selection of how to split the dataset at each node. This involves selecting the feature and, where applicable, the threshold or class that best splits the data into homogeneous subsets to enhance predictive capability. Impurity measures offer the metrics for evaluating potential splits, guiding the tree-building algorithm toward good splits.

**Entropy** from information theory is a quantification of the disorder or uncertainty in a dataset. For classification problems, it is a quantification of the mixing of the classes in a node. Entropy of a dataset S that has n classes is mathematically expressed as:

$$Entropy(S) = \sum_{i=1}^{n} p_i \log_2(p_i) \tag{1.4}$$

where $p_i$ is the fraction of data points that are in class i. Entropy is zero when all data points are in one class (representing complete purity). Entropy is at a maximum when classes are balanced, representing maximum uncertainty.

**Information Gain** (IG) quantifies how effectively a feature divides the data by quantifying the reduction in entropy from a parent node to its child nodes as a result of splitting. It is defined as:

$$IG(S,A) = Entropy(S) - \sum_{v \varepsilon Values(A)} \frac{|S_v|}{|S|} Entropy(S_v) \tag{1.5}$$

where S is the original dataset, $S_v$ is the subset of S where feature A has value v, and $|S_v|$, $|S|$ are their sizes. Features with the highest information gain are selected for splitting, as they lead to the maximum increase in node purity.

**Gini impurity**, as prominently used in the **CART algorithm**, is another measure of node impurity [55]. It is the chance of misclassifying an element chosen randomly if it were labeled with a class label according to the label distribution in the node:

$$Gini(S) = 1 - \sum_{i=1}^{n} p_i^2 \tag{1.6}$$

Gini impurity is zero when all instances belong to one class and reaches its maximum when the classes are uniformly distributed. Compared to entropy, Gini impurity is computationally less intensive and often produces comparable splits.

Decision tree algorithms evaluate the impurity measure (entropy or Gini) at each candidate split and select the one that maximizes impurity reduction, effectively improving **homogeneity of the resulting nodes** [56]. In the case of regression tasks, impurity is commonly measured using metrics such as **mean squared error (MSE)** or **mean absolute error (MAE)**. By recursively applying splitting criteria in this manner, decision trees iteratively partition the feature space in a way that maximizes predictive capability while maintaining interpretability. Uncontrolled growth, however, can cause the tree to overfit, learning noise rather than general trends [57]. This is addressed by pruning and regularization techniques discussed later.

Decision trees are grown using recursive partitioning, which splits the dataset into more homogeneous subsets at every step. The algorithm selects the optimal feature and split value to divide the data at every step in a manner that maximizes class purity for classification problems or reduces variance for regression problems. Decision trees use impurity measures to determine the optimal splits. Both IG and Gini impurity favor features that create the most distinctive class separations. Decision trees can work with both categorical and continuous features. The categorical values are split by dividing categories, and continuous values are split at the optimal thresholds discovered during training. This renders decision trees viable for a wide variety of data types, eliminating the need for preprocessing such as normalization.

Decision tree learning has developed around a series of landmark algorithms that have each introduced methodological advancements in learning accurate and interpretable models. The most notable of these are ID3, C4.5, and CART, and many decision tree implementations today are based on these.

Introduced by Quinlan in 1986, **ID3** was one of the first algorithms to formally apply information theory for the construction of decision trees. It operates by selecting, at each node, the attribute that provides the most information gain — a measure of how well a feature separates the training data into pure classes [53]. The algorithm recursively partitions the data set based on this criterion, creating a tree that classifies instances by following a path of feature-based decisions. ID3 only supports categorical attributes and does not perform pruning, and is thus prone to overfitting.

**C4.5** is an extension to ID3 that addresses some of its shortcomings. Notably, it introduces the gain ratio, which normalizes information gain by the intrinsic information of a split, to avoid bias toward attributes with many levels. C4.5 also accommodates continuous features by dynamically choosing threshold splits, and it can handle missing values during both training and prediction. Furthermore, the algorithm incorporates post-pruning methods to prevent overfitting, thereby improving generalization to unseen data. These enhancements have made C4.5 a very widely used algorithm in machine learning and data mining.

Introduced by Breiman (1984), **CART** is defined by the binary splits, a difference point where every node divides the data into exactly two branches based on a threshold condition. In classification, CART utilizes the Gini impurity measure to assess the quality of splits, and for regression, the minimization of mean squared error (MSE). Unlike C4.5, which can generate multi-way splits, CART will always construct a strictly binary tree. CART also features aggressive cost-complexity pruning, wherein branches that don't significantly impact prediction are pruned to reduce the complexity of the model. These algorithms formed the basis of modern tree-based methods and serve as a foundation for altering and optimizing decision trees for a range of supervised learning tasks (Table 1.2).

Decision trees **overfit**, especially when they grow very large and match the noise in the training data. Pruning techniques must be used to simplify these models, enhancing their ability to generalize to new, unseen data. There are two significant pruning strategies: pre-pruning and post-pruning, with various techniques and implications.

**Pre-pruning** prevents the decision tree from growing when it is still being constructed. It does so by setting some limits, not letting the tree grow overly complex. Some pre-pruning techniques include:

- Maximum depth limitation: Restricting the depth of the tree so that it will not grow too deep.
- Minimum samples per node: Stipulating the minimum samples required to split a node.
- Minimum information gain: Requiring at least some minimum increase in information gain for a split to be made.

Cheap to compute, pre-pruning runs the risk of underfitting, since the model may not be able to capture all the significant patterns in the data [57].

**Post-pruning** involves allowing the decision tree to grow to its fullest depth and then systematically pruning branches that contribute little to its predictive ability. This method achieves better generalization performance than pre-pruning. One of the most well-known post-pruning methods is cost-complexity pruning, or weakest link pruning. It balances the complexity of the tree with its accuracy by optimizing the cost function:

**Table 1.2** Comparison among ID3, C4.5 and CART

| Feature | ID3 | C4.5 | CART |
|---|---|---|---|
| Splitting criterion | Information gain (based on entropy) | Gain ratio (addresses bias toward many-valued attributes) | Gini impurity or entropy |
| Attribute types | Categorical only | Handles both categorical and continuous attributes | Handles both categorical and continuous attributes |
| Handling continuous data | Not supported directly | Splits continuous attributes by selecting threshold values | Splits continuous attributes by selecting threshold values |
| Pruning method | None (pure growth) | Post-pruning using pessimistic error estimation | Cost-complexity pruning |
| Missing values | Not handled | Supported (probabilistic assignment of samples) | Supported (surrogate splits) |
| Output structure | Multi-way decision tree | Multi-way decision tree | Binary decision tree |
| Typical use cases | Simple concept learning, small categorical datasets | General-purpose decision trees with mixed attribute types | Classification and regression tasks |

$$R_\alpha(T) = R(T) + \alpha|leaves(T)| \tag{1.7}$$

where:

- R(T): Misclassification rate of tree T
- |leaves(T)|: Number of leaf nodes of tree T
- α: Complexity parameter penalizing the number of leaves

Decision trees have a number of strengths, which have rendered them very popular both for research and application-oriented work in intelligent systems. Perhaps the most significant strength is interpretability. Each decision node corresponds to a logical test on a feature, and the path from the root to the leaf effectively specifies a clear, rule-based classification or regression logic. This transparency is particularly welcome in domains where model decisions must be explainable, such as financial risk modeling or medical diagnostics.

Another benefit is the absence of lengthy data preprocessing. Decision trees do not require linearity and do not necessitate normalization or feature scaling, making them highly suitable for heterogeneous datasets. Additionally, they can naturally handle both categorical and numeric data, and extensions of the basic algorithms, such as C4.5 or CART, can handle missing values and continuous attributes (Fig. 1.14).

Decision trees also suffer from serious drawbacks, however. One of the primary drawbacks is that they can overfit, especially if the tree is allowed to grow unchecked. Overfitting occurs when the model picks up noise in the training data, reducing its

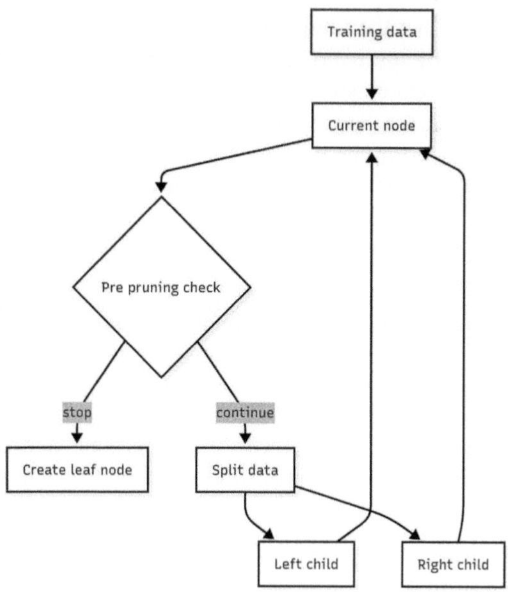


**Fig. 1.14** The concept of pre-pruning a decision tree

capacity to generalize to new data. It is for this reason that techniques like pruning and ensemble techniques (e.g., Random Forests) are often applied to control complexity. Another limitation is model instability where small changes in the data used for training can result in drastically different tree structures. Such sensitivity compromises their robustness, particularly if they are employed as standalone predictors. Besides, although trees are simple in structure, they can become unwieldy with high-dimensional feature spaces or deep hierarchies, which results in loss of interpretability in practice. Despite these weaknesses, decision trees remain at the heart of machine learning, as they are the building blocks of a vast array of advanced ensemble methods that remedy their weaknesses but retain their strengths.

Decision trees are used extensively in numerous fields due to their interpretability and flexibility. In medicine, they facilitate clinical decision-making by aiding in diagnosis, prognosis, and treatment planning, providing comprehensible rules that are interpretable and verifiable by clinicians. In finance, decision trees facilitate credit scoring, fraud detection, and risk assessment, where transparent decision-making paths are essential for regulatory compliance and effective risk management. Marketing utilizes decision trees for customer segmentation, churn prediction, and personalized recommendation systems, leveraging their ability to handle heterogeneous data and yield compelling results.

### 1.2.5 Support Vector Machines

**Support Vector Machines (SVMs)** are powerful supervised algorithms used primarily for classification, but they can also perform regression and outlier detection. Briefly, SVMs operate by discovering the best **hyperplane** that maximally separates data points of different classes in a high-dimensional space [58]. The optimal hyperplane is the one that achieves the largest margin between the two classes, where the margin is defined as the distance between the hyperplane and the nearest data points from each class, termed support vectors.

$$D = \{(x_1, y_1), \ldots, (x_n, y_n)\}, \quad \text{where } x_i \in R^d \text{ and } y_i \in \{-1, 1\} \tag{1.8}$$

A linear SWM seeks a hyperplane defined by the equation:

$$w^T x + b = 0 \tag{1.9}$$

where w is an orthogonal weight vector to the hyperplane and b is the bias. The objective is to maximize the margin under the constraint that all training data satisfy:

$$y_i \left(w^T x_i + b\right) \geq 1 \forall i \tag{1.10}$$

The optimal solution is obtained by solving a convex quadratic optimization problem, thus ensuring both computational efficiency and global optimality. SVM establishes a decision boundary not only to classify the data but also to generalize well to new examples. This is achieved by ensuring that the classifier is kept as far as possible from the training set boundaries, thereby minimizing the risk of overfitting.

**Linear SVMs** are suited where data can be separated by a straight hyperplane or line. The models attempt to maximize the margin, the distance of the nearest data points of each class (support vectors) to the hyperplane. Maximum margin is created to enhance model generalization and resilience. Mathematical representation of the decision boundary is:

$$f(x) = w'x + b \tag{1.11}$$

where w is the weight vector that is orthogonal to the hyperplane, and b is the bias. When there's perfect separability, the SVM computes the hyperplane with minimum misclassification.

Most real-world data sets exhibit non-linear relationships that cannot be accurately approximated by a linear hyperplane in the original input space. Support Vector Machines (SVMs) achieve this problem through a method referred to as the kernel trick, where the model is able to implicitly map the input features to a higher dimension without ever actually performing the mapping [59]. This enables the algorithm to learn the linear separability in the higher-dimensional space, which creates a non-linear decision boundary in the input space. Instead of explicitly computing the transformation φ(x), the kernel trick computes the dot product ⟨φ($x_i$), φ($x_j$)⟩ by means of a kernel function K ($x_i$, $x_j$). Not only does this maintain computational efficiency but also enables SVMs to tackle intricate classification tasks in fields such as handwritten character recognition, genomic sequence comparison, and fraud detection (Fig. 1.15).

**Benefits of the Kernel Approach**

- Dimensional Sscalability: Kernels can efficiently simulate mappings to infinite-dimensional feature spaces (e.g., RBF kernel) with virtually zero computational cost.

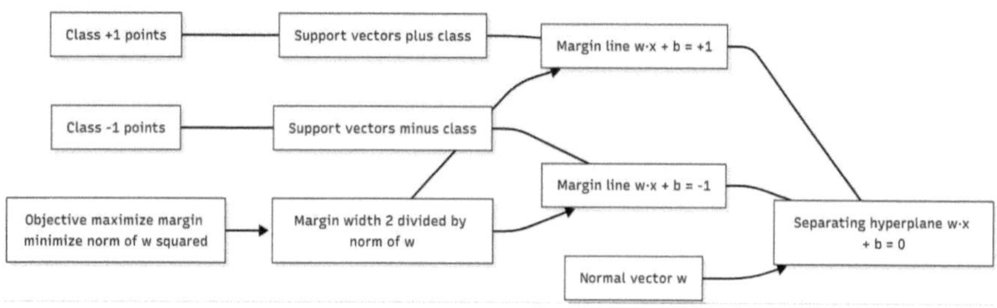


**Fig. 1.15** Diagram of a separating hyperplane and margin in an SVM classifier

- Wide applicability: The technique is not only used in SVMs, but also in Kernel PCA, Kernel Ridge Regression, and numerous other algorithms.
- Domain-specific design: Kernels can be specially tailored to adapt to domain-specific data structures, e.g., string kernels in bioinformatics or graph kernels in cheminformatics.

For the kernel trick to be implemented, different kernel functions are utilized based on the type of dataset. Each kernel maps data into a space where linear separability is possible.

Support Vector Machines (SVMs) are commonly applied across various areas, including image recognition, bioinformatics, and text classification, as they can effectively handle high-dimensional data. Their ability to perform well in feature spaces with higher dimensionality than the sample size makes them optimal for complex applications such as genomic data processing and natural language processing [60]. The resistance to overfitting of SVMs is a consequence of their margin maximization strategy and the presence of regularization parameters, making them robust despite limited training data. SVMs are not perfect either; they are computationally costly to train and memory-intensive, especially when nonlinear kernels are used on large data sets. Furthermore, unlike decision trees, the complex decision boundaries of SVMs resulting from kernel functions often reduce the interpretability of the models, which is a challenge in applications where model transparency is paramount. All these notwithstanding, SVMs remain a robust and pervasive component of current intelligent systems, inciting ongoing research in scalable and hybrid learning methods (Table 1.3).

### *1.2.6 Model Evaluation Matrix*

Evaluating the performance of classification models is a critical aspect of the machine learning workflow since it indicates how well a model is able to make correct predictions on fresh, unseen data. Ultimately, model evaluation is intended to ensure that the classifier generalizes well from the training set, neither overfitting nor underfitting. To achieve this, various metrics have been developed, each of which extracts different aspects of model performance depending on problem type and data form. Among the most commonly used metrics are accuracy, which holds the number of correctly classified instances against the number of predictions made. Accuracy on its own can be misleading, however, especially when considering imbalanced datasets, i.e., classes do not occur with equal frequency. For example, in fraud detection or disease diagnosis, the minority class is often of utmost significance, and accuracy-maximized models will overlook the rare but highly significant cases [11]. To better quantify performance on imbalanced datasets, measures such as precision and recall are utilized. Precision is concerned with the accuracy of correct positive predictions, measuring how many correctly predicted positives are

**Table 1.3** Comparison of Common Kernel Functions Used in SVMs

| Kernel | Mathematical Form | Key Parameters | Characteristics | Typical Use Cases |
|---|---|---|---|---|
| Linear | $K(x,x') = x^{T}x'$ | None (or regularization C only) | Fast to compute; works well when data are approximately linearly separable; low risk of overfitting | Text classification, high-dimensional sparse data |
| Polynomial | $K(x,x') = (\gamma x^{T}x' + r)^{d}$ | Degree d, coefficient r, scale γ | Captures interactions up to order dd; flexible but computationally heavier for high degrees | Image recognition, problems with moderate feature interactions |
| Radial basis function (RBF/ Gaussian) | $K(x,x') = \exp.(-\gamma\|x - x'\|^{2})$ | Gamma γ | Very flexible; maps to infinite-dimensional space; good default choice but sensitive to parameter tuning | Non-linear decision boundaries, most general tasks |
| Sigmoid | $K(x,x') = \tanh(\gamma x^{T}x' + r)$ | Scale γ, coefficient r | Behaves like a two-layer neural network; can be tricky to tune; not always positive-definite | Certain neural-network-inspired applications |

indeed relevant, whereas recall measures how well the model can recognize all relevant positive instances within the data.

The trade-off between precision and recall is typically represented by the F1 score, which is the harmonic mean of the two and provides a single measure that is equally sensitive to both false positives and false negatives. Further insight into classifier performance is provided by the confusion matrix, which classifies predictions into true positives, false positives, **true negatives**, and **false negatives**. This matrix also serves as the basis for the majority of performance measures, providing a clear picture of where the model is performing well and where it is not. In addition to scalar measures, graphical tools such as the **Receiver Operating Characteristic (ROC) curve** and **Precision-Recall (PR) curve** provide a visual representation of performance as decision thresholds change. The ROC graph plots the false positive rate against the true positive rate, and the **area under the curve (AUC)** will reflect the overall discriminative power.

For datasets that are **class-imbalanced**, PR curves are normally employed instead since they will yield a more explicit measurement of the precision-recall trade-off of the positive class. Together, these measures and visualizations constitute a comprehensive set of tools for model classification. The choice of appropriate measures depends on the specific application, data distribution, and the comparative costs of false positives and false negatives.

**Accuracy** is the most straightforward and frequent test metric for classification task evaluations. It is defined as the ratio of the number of instances accurately predicted (positive and negative) to the total number of predictions.

$$Accuracy = {(TP + TN)}\Big/{(\mathrm{TP} + TN + FP + FN)} \tag{1.12}$$

where:

- TP = True Positives (positive instances correctly predicted)
- TN = True Negatives (negative instances correctly predicted)
- FP = False Positives (predicted as positive but are not)
- FN = False Negatives (predicted as negative but are not)

This measure gives us a rapid glimpse of the overall performance of a model, particularly in the case of balanced datasets when the classes are uniformly distributed. Its efficacy diminishes when there is a class imbalance.

For example, in a medical diagnosis situation where just 1% of the population suffers from a specific disease, a classifier that classifies all patients as healthy will remain 99% accurate—yet miss all actual cases, making it useless for the desired application. As this is the main flaw, accuracy must be supplemented with other measures such as precision, recall, or F1 score while judging classifiers on imbalanced datasets.

In classification problems of imbalanced class distribution, the usual measures like accuracy fail to properly represent model reliability for the minority class. Precision and recall thus provide more refined perspectives by focusing on positive predictions only.

**Precision** calculates the proportion of instances it labels as positive that are in fact positive. High precision indicates that the model has very few false positives and is extremely useful in cases such as spam filtering or medical alert systems, where false alarms are too costly or dangerous. It is calculated by the formula:

$$Precision = TP\Big/(TP + FP) \tag{1.13}$$

**Recall**, also referred to as **sensitivity** or the **positive actual rate**, is the ratio of true positive cases correctly reported by the model. Having a high value for recall ensures that most of the positive instances are discovered. It is a significant attribute in use cases such as cancer detection or fraud detection, where missing a positive case (false negative) can have disastrous consequences.

$$Recall = TP\Big/(TP + FN) \tag{1.14}$$

Precision and recall are usually inversely proportional: when one increases, the other decreases based on the classification threshold. The model can be adapted for use with different levels of the same threshold. For instance, in high-risk medical diagnosis, recall optimization is more important than precision, as it attempts to capture all cases at the cost of reducing precision.

In classification issues, particularly class imbalanced distribution ones, the limitations of using accuracy are apparent. While precision emphasizes the accuracy of positive predictions and recall emphasizes the ability to find all instances of note, neither is the whole story by itself. The **F1 score** is intended to address this by combining both measures into a single measure of performance. The mathematical formulation for F1 score is as the harmonic mean of precision and recall:

$$F1\,score = \frac{2\times(Precision \times Recall)}{(Precision + Recall)} \tag{1.15}$$

This definition is one where the F1 Score is only beneficial if both precision and recall are satisfactory. It also penalizes extreme imbalances between the two, which makes it well-suited for domains where both false positives and false negatives are of paramount importance, such as in medical diagnosis, cybersecurity, or fraud detection.

A **confusion matrix** is a key model classification evaluation tool that provides extensive insight into how well a model is performing by relating predicted to actual classifications. It is most commonly used when assessing binary classifiers, but it can also be applied to multiclass issues. The matrix exists in the form of an ability in binary classification problems, showing the amounts of true positives (TP), true negatives (TN), false positives (FP), and false negatives (FN).

This structure enables the computation of several key performance metrics, including accuracy, precision, recall, and the F1 score. More importantly, the confusion matrix reveals specific types of mistakes a model will make, a consideration relevant in cases of non-uniform misclassification costs [26]. For example, in medical screening, a false negative (failing to detect a disease) could have more severe consequences than a false positive.

The confusion matrix also enables **class-specific measurements** like the **False Positive Rate (FPR)** and **True Negative Rate (TNR)**, which provide further information regarding classifier performance across different thresholds. It may be employed in conjunction with graphical tools like the ROC curve, which plots TPR vs FPR graphically.

In practice, confusion matrices are indispensable for the identification of issues in classification systems and for steering improvements. For example, a matrix imbalance can indicate that class imbalance needs to be addressed using methods such as resampling, cost-sensitive learning, or adjusting the decision threshold (Fig. 1.16).

The **ROC curve** is a performance evaluation tool that allows us to assess a model's discriminatory ability at various decision thresholds. This is especially useful

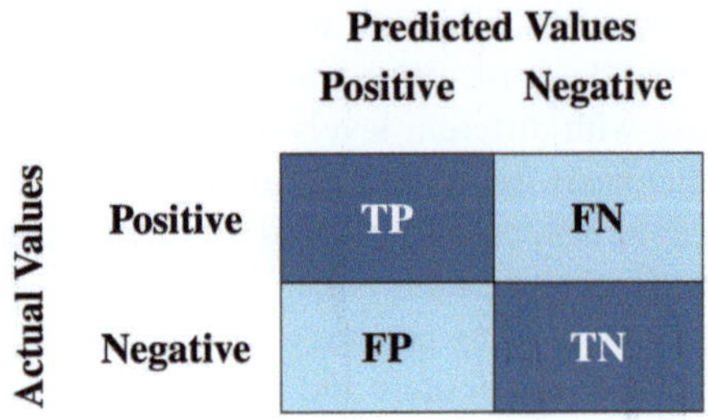


**Fig. 1.16** Confusion Matrix Diagram (https://commons.wikimedia.org/wiki/File:Confusion_matrix;_sep_2025.png)

when different thresholds result in different trade-offs between correct and incorrect classification of positive and negative examples.

$$TPR = {TP}/{(TP + FN)} \tag{1.16}$$

$$FPR = {FP}/{(FP + TN)} \tag{1.17}$$

A model with ideal predictions will yield a point at the top-left of the ROC space (TPR = 1, FPR = 0). A model providing random predictions will produce a diagonal line from the bottom-left to the top-right where TPR is equal to FPR at all thresholds. The nearer the curve is to the top-left corner, the more accurate the model is over thresholds.

**AUC** gives a single scalar value measuring a classifier's overall performance. It is an estimation of the probability that the classifier will rank a randomly chosen positive instance higher than a randomly chosen negative instance.

$$AUC = \int_0^1 TPR(x)\,dx \tag{1.18}$$

Interpretation of AUC values:

- AUC = 1.0: Perfect classifier
- AUC ≥ 0.9: Excellent discrimination
- 0.7 ≤ AUC < 0.9: Good to acceptable
- 0.5 ≤ AUC < 0.7: Slightly better than random, but poor
- AUC = 0.5: No discriminative power

This threshold-independent measure reduces the ROC curve to a single number and is therefore particularly handy for model comparison. The ROC-AUC metric has broad applicability across areas where class discrimination is more important than optimizing overall accuracy. For instance, in disease diagnosis in medicine, high AUC values indicate that disease identification is reliable. Consider the following:

- In fraud detection, AUC is utilized to balance identifying fraudulent transactions (TPR) and reducing false alarms (FPR).

- In machine learning competitions, AUC is utilized as a standard measure owing to its insensitivity to various thresholds.

In addition, ROC curves can be utilized to:

- Compare various classifiers on the same test set.
- Choose an optimal threshold based on cost-sensitive tasks.
- Detect overfitting or underperformance based on the shape of the curve.

Even though ROC curves are robust, they are misleading when the data is highly imbalanced. There, the False Positive Rate can unrealistically be low just because there are so many true negatives, and the performance will seem good. This renders ROC less helpful when the negative class significantly outnumbers the positive class. Precision-Recall curves are preferable in such a scenario.

Choosing a classification algorithm involves compromises across a number of dimensions, **interpretability, generalization, scalability**, and **computational efficiency**. Decision Trees and SVMs are two widely used but highly disparate algorithms in terms of design and usage. Their relative comparison decides the optimal method based on data structure, noise levels, feature types, and target application areas.

Decision Trees are highly interpretable. The tree structure provides simple rules for each decision path, and the model is thereby interpretable to even a layperson. This renders them extremely useful in high-stakes fields like healthcare, finance, and law, where model transparency is required for accountability and regulatory compliance. SVMs, by contrast, especially those with non-linear kernels, are black-box models. Their decision boundaries are computationally complex and lack an intuitive explanation, making their application challenging in scenarios demanding transparency.

SVMs are inherently suited to non-linearly separable data through the kernel trick, with mapping to high-dimensional feature spaces where linear separation is possible [39]. This allows for modeling intricate decision boundaries, such as concentric class separations or XOR-type distributions. Decision Trees, while flexible through ensemble techniques such as Random Forests and Gradient Boosting Machines, have a greedy approach to divide the feature space and may require post-processing (e.g., pruning) to control complexity.

Decision Trees are very sensitive to noise, and small alterations in the data can result in radical differences in tree structure. This instability has a tendency to result in overfitting, especially if the model is grown deep and not pruned. SVMs, on the other hand, maximize the margin between classes, which gives immunity from overfitting, especially with regularization techniques. However, they are undermined by class imbalance unless specialized methods, such as cost-sensitive learning, class weighting, or resampling techniques, are utilized.

Decision Trees are trainable fairly fast due to their recursive partitioning strategy. Their performance scales well with moderately large datasets and receives a further boost when executed in distributed settings. SVMs, on the other hand, require the solution of a quadratic optimization problem, which is computationally prohibitive for large-scale datasets, especially in non-linear cases. Approximation methods, such as Sequential Minimal Optimization (SMO) or linear SVM versions, improve scalability but do not entirely eliminate the computational barrier.

Decision Trees natively handle categorical and continuous features without preprocessing. They are invariant to feature scales and outlier-resistant due to their hierarchical structure. SVMs, in contrast, demand careful preprocessing: input features must be scaled or normalized, and categorical variables must be one-hot or ordinal encoded. This preprocessing can introduce complexity and sources of bias when working with high-dimensional or heterogeneous datasets.

Decision Trees are applied extensively in applications wherein interpretability and logical decision-making are significant. These include, for instance, medical diagnosis (e.g., risk scoring), business analysis, and policy rule modeling. SVMs are popular in applications wherein high accuracy and computational intensity are required, such as image classification, speech classification, handwriting recognition, text classification, and bioinformatics. They perform well on problems with high-dimensional input spaces and complex boundary conditions.

## 1.3 Advanced Supervised Learning

One of the machine learning algorithm techniques is **supervised learning**. It uses human-tagged input and output data for training artificial intelligence models. The trained algorithm learns the relationships inherent in the inputs and the outputs and can predict the right outputs for new and unlabeled real-world input data. Labeled data includes example points along with the right responses or outputs. Given the input data in the machine learning algorithm, it will refine the weights until the model fits just right. The labeled training data instructs the model how to spot the relationships and what kind of patterns to search for between the aspects and the labels of the data.

Supervised learning allows one to solve various real-world tasks, such as the identification of spam or the forecasting of the stock price. It can also contribute to the creation of very accurate machine learning models. **Advanced supervised learning** is a newer and more sophisticated set of methods devised for the task of improving the performance, reliability, and interpretability of models for supervised or labelled learning. In supervised learning, one attempts to get a mapping function and transform the inputs in terms of known outputs by the use of training data [61, 62]. Advanced supervised learning does more than this and also incorporates **ensemble techniques** such as **bagging, boosting, and stacking.**

## 1.3.1 Ensemble Learning

**Ensemble learning** is a method of machine learning in which two or more learners are integrated, such as regression or neural networks, in an endeavor to create better predictions. An ensemble model integrates a set of individual models in the production of more precise predictions than a solo model can yield individually [63]. This approach is founded on the belief that a group of learners offers overall improved accuracy compared to a solo learner, and is also supported by research. It is also desirable in the sense that in the event one learner is in error, others can auto-correct them, and the resultant model is less inclined towards a state of overfitting, more resilient, and unlikely to be affected by minor variations in the training set (Fig. 1.17).

The ensemble learning models are either parallel or sequential. The parallel procedures borrow their basis from the training of the individual base learner or the weak learner, individually from the rest. In this approach, various models are trained autonomously, simultaneously in parallel, typically under different subsets of the training samples. The predictions of the models are then combined by voting or averaging in the final prediction during training. The parallel ensemble procedures are primarily used for variance reduction in machine learning models, particularly for unstable models.

Parallel approaches can also be classified as homogeneous and heterogeneous approaches. The homogenous parallel ensembles employ the same general learning algorithm to generate the entire set of component base learners, whereas the heterogeneous parallel ensembles employ various algorithms in generating base learners. **Sequential ensemble learning** is a method in which a progression of base models or weak learners is trained individually in such a manner that the next one is influenced by the errors in the prior one. In this manner, a chain of learners is established in which, at every step, the errors or faults in the earlier steps' models are corrected [64, 65]. The type of method is a method of trying to **reduce bias** by iteratively improving performance in such a manner that the next models try and address the errors of the prior models [66] (Fig. 1.18).

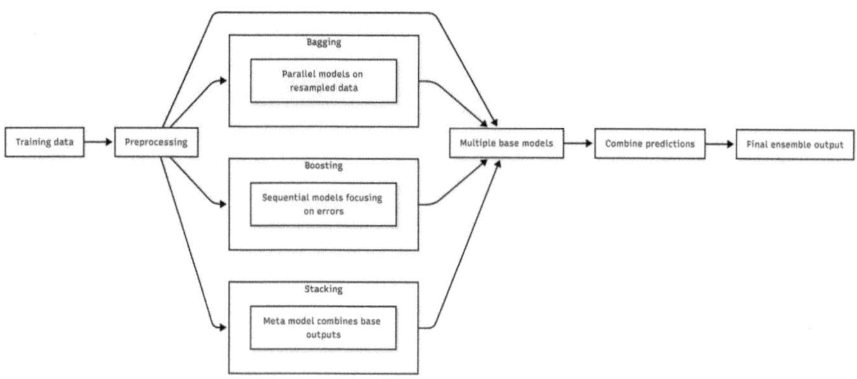


**Fig. 1.17** Ensemble learning principle

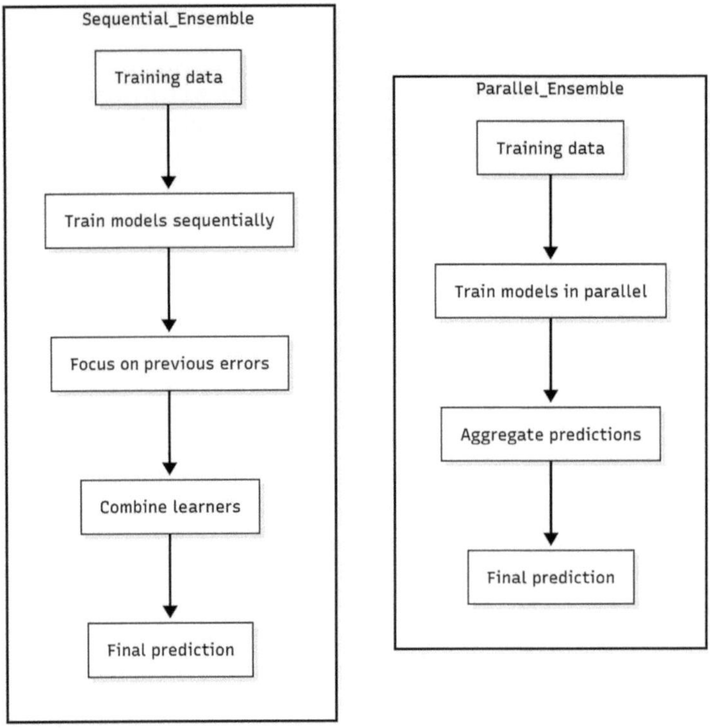


**Fig. 1.18** Parallel vs. Sequential Ensembles

Most of the ensemble methods combine base or weak learners in a final learner by a technique known as **stacking**, in which a separate machine learning algorithm trains an ensemble learner from the base learner. But one of the very common techniques for assembling the predictions of base learners is majority voting. Each base learner's estimate for a given point is considered by this approach, and gives a final estimate based on what the majority of the learners estimate.

**Majority voting** is one of the most common ensemble techniques in a binary classification problem in which various base or weak learners classify the same set of data. The decision or final result is a straightforward majority vote, i.e., whatever the class the majority of the classifiers are voting for becomes the ensemble's output. For instance, assume three classifiers have provided the following classes [1, 0, 1]. The final prediction will then be class 1 because it was voted for twice. Weighted majority voting is a continuation of this approach. Rather than voting equally for each classifier, weights are given for them based upon reliability or past performance, for example, accuracy in a validation set. The highly weighted classifiers give more weight in the last decision. For example, if classifier A performs better than classifiers B and C, a larger weightage (A:0.5, B:0.3, C:0.2) can be given for them. While voting, the outputs of these classifiers are scaled by their weights, and the class of the output having the maximum weighted sum is chosen as the final output.

This method is particularly useful for heterogeneous sets, wherein different types of models are combined, and a few of them are more competent at a task than others [67]. In utilizing the best of their own abilities, weighted voting tends to boost aggregate model competency and robustness.

Bagging, boosting, and stacking collectively embody differences between sequential and parallel, as well as homogeneous and heterogeneous, forms of ensemble approaches, and represent various strategies for combining one or more base learners to enhance predictive accuracy and generalization.

**Bagging** is also referred to as **bootstrap aggregating**, for short. It is a homogeneous and parallel ensemble method. It is a homogeneous method in the fact that the base learners or the models within the ensemble are similar in kind. While the name parallel implies the fact that the models are separately and simultaneously trained without any interaction. It therefore trains a number of base learners using the same training algorithm by leveraging the replicates of a given training set.

The primary objective of this approach is the minimization of variance and the enhancement of stability by training a base learner individually using a random sample of the training set. The sampling is without replacement. This is referred to as bootstrapping. We can assert that a training set comprises n training instances. The bootstrap resampling replicates n instances from the set into a new subsample set, while the first instances appear more than once, and others not at all [68, 69]. These are referred to as bootstrap samples. Recurring the same process x times will yield x iterations of the same primary set, and these iterations consist of n samples originated from the primary set. The iteration of the primary set is then employed in training a distinct base learner using the same learning algorithm (Fig. 1.19).

**Stacking** or **stacked generalization** is a heterogeneous parallel ensemble learning technique that operates by compiling predictions of a set of base learners in order to enhance overall performance. The heterogeneity implies that the base learners are various forms of models, and this heterogeneity helps take advantage of the distinct strengths of different algorithms, making the method overall more robust. When we also consider it as a parallel method, it implies that the base models are all individually trained at the same time. Stacking is a classic example of meta-learning, which is the process of training a meta-learner from the outputs of numerous base learners [70]. Stacking specifically trains a few base learners from

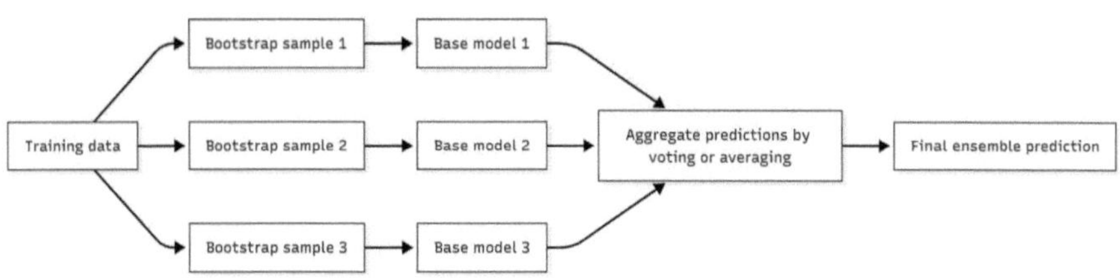


**Fig. 1.19** Bagging ensemble principle

the same set in a different algorithm for training each learner. The base learners make predictions about an unseen set. The predictions of the first model are then collected, and a final model is trained from them, i.e., the meta-model.

It is highly necessary to employ a different set for training the meta-learner, rather than the set employed for training the **base learners**, as training the base and the **meta-learners** using the same set of data would lead to overfitting. The meta-learner may memorize the base-learners' predictions rather than learning how to generalize, and in order to prevent it from occurring, a part of the training set should be reserved or not used for the training of the base-learners. Instead, this reserved set is employed for making predictions by the base-learners, and such predictions are given as inputs for the training of the meta-learner. One widely used method for the adequate stratification of the data is cross-validation. The use of the k-fold cross-validation in the context of the stacking method provides the assurance that any point from the data is employed for the training and validation set, but not concurrently, and hence the meta-learner receives non-overlapping and reliable training data [71]. By doing this, we ensure that the meta-learner does not simply rely on the base learners, but rather learns their patterns and therefore generalizes the novel data better (Fig. 1.20).

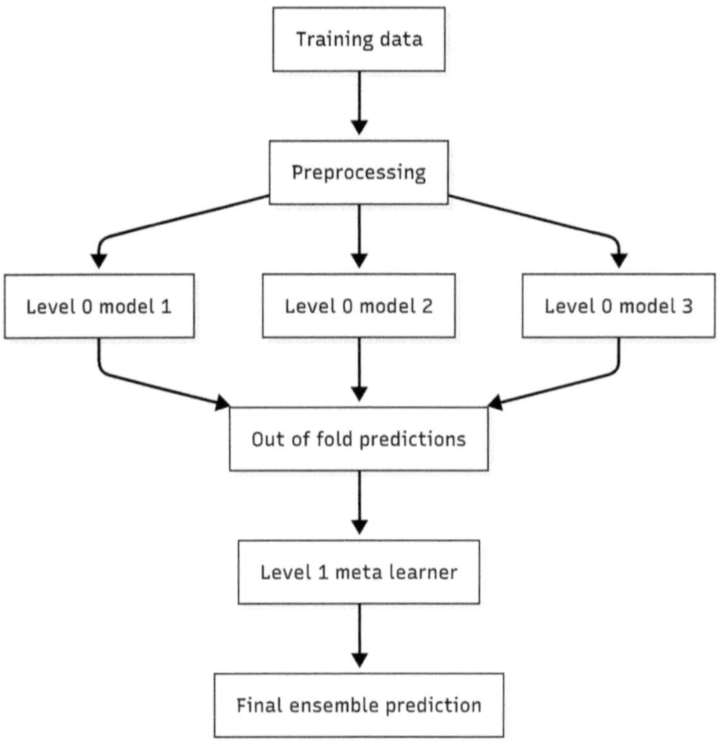


**Fig. 1.20** Principle of stack ensembles

**Boosting** is a sequential ensemble learning technique used to build strong predictions or predictive models by combining multiple weak or base learners, trained in a sequential manner. Boosting has many variations, all of which follow the same general procedure. The process begins with training of the weak or base learners on the first or original dataset, denoted as d1. Since this learner is weak, it will likely classify many data instances incorrectly. Instead of randomly sampling data, as in bagging, boosting focuses on the cases that were misclassified by the previous model. A new dataset, d2, is created by assigning higher weights or importance to those misclassified examples, increasing their chance of being correctly learned in the next round. A second learner is trained on d2. Again, its performance is evaluated, and a new dataset d3 is generated, which focuses on the errors made by the second learner and potentially on examples where the first and second learners disagree [72]. This process is repeated for a fixed number of iterations (n) or until acceptable performance is attained. All new learners are trained to learn from previous learners and correct the errors they made. After training all n learners, boosting combines them to make the final prediction. The contribution of each model is typically weighted based on its accuracy, meaning better-performing models have more influence on the final result.

Boosting algorithms can be classified based on how they prioritize erroneously predicted data instances when creating a new dataset, with two main approaches being **Adaptive Boosting** and **Gradient Boosting**. Adaptive Boosting, also known as **AdaBoost**, works by weighting model errors. When creating a new iteration of a dataset for training the next learner, AdaBoost adds weights to the previous learner's misclassified samples, causing the next learner to prioritize those misclassified samples. Gradient Boosting uses residual errors while training new learners [73, 74]. Instead of utilizing the weight of misclassified samples, gradient boosting uses residual errors in the previous model to set target predictions for the next model. It therefore attempts to fill the gap in error created by one model.

### *1.3.2 Random Forests*

**Random Forest** is a learning algorithm that utilizes decision trees in making more accurate predictions. Different random samples of the data are employed by various trees and voting for classification, and averaging for regression, in compiling their outputs in a way that yields more accurate predictions and fewer mistakes. A decision tree is a classification and regression supervised learning algorithm and a core component of random forests. It operates by representing a series of decisions in a tree-like framework wherein the internal node is a test applied by a feature and the branch is the test result, and a single leaf is a predicted final output. A decision tree simulates human decision-making by decomposing issues in the simplest possible way in terms of yes/no questions and resulting in a final prediction.

The Random Forest algorithm starts by generating numerous decision trees. The trees are trained in smaller subsets of the primary set of data using bootstrap sampling, such that the data is chosen without replacement. Consequently, a different version of the training set is seen by each tree [75, 76]. When constructing a tree, the algorithm does not use a full set of existing features at each split. Instead, the algorithm uses smaller and smaller subsets of the existing set of features randomly and picks the best split. Consequently, the trees differ from one another and thereby decrease the correlation between the trees. After constructing the entire set of trees, individual trees can make predictions based on what they have learned by constructing their own set of data, and are also called base learners. The two primary methods for combining predictions are classification and regression. In classification, a class label is voted on by the individual tree, and the final output is the majority vote or the class decided by the majority of the trees [77]. While in regression, a numerical value is predicted by the individual tree, and the final output is the average of the predictions. Since this algorithm uses random subsets of the data and also incorporates diversity in the trees, the potential for overfitting is reduced and the model's generalization is enhanced for new and unseen data [78]. Since averaging predictions of many different trees, random forests generate more stable and precise predictions than an individual decision tree does.

### *1.3.3 Gradient Boosting Machines*

**Gradient Boosting Machines** are programs and machines that employ the use of ensemble procedures by basing the functions upon them. It implies that the procedures use randomization for the creation of a few different solutions to one problem at hand. These different variants of solutions enable the machine to train at a faster pace and also improve generalization. The boosting algorithms utilize various weak learners and aggregate them into a single strong learner, operating in an iterative process [79]. That is why gradient boosting machine learning is indispensable in applications such as predicting bioactive molecules and identifying dementia from electronic health records, it effectively and accurately handles high volumes of intricate tabular data.

**XGBoost** is a gradient boosting decision tree algorithm operating under high scalability in the construction of an additive expansion in the objective of a task at hand. The key characteristic of such systems is the loss minimization or regularization by varying the depth of the tree. It uses a variation of a loss function for this task:

$$\Omega(h) = \gamma T + 1/2\lambda \|w\|^2 \tag{1.19}$$

where T is the number of leaves in the tree and w is the leaf's outputs (Figs. 1.21 and 1.22).

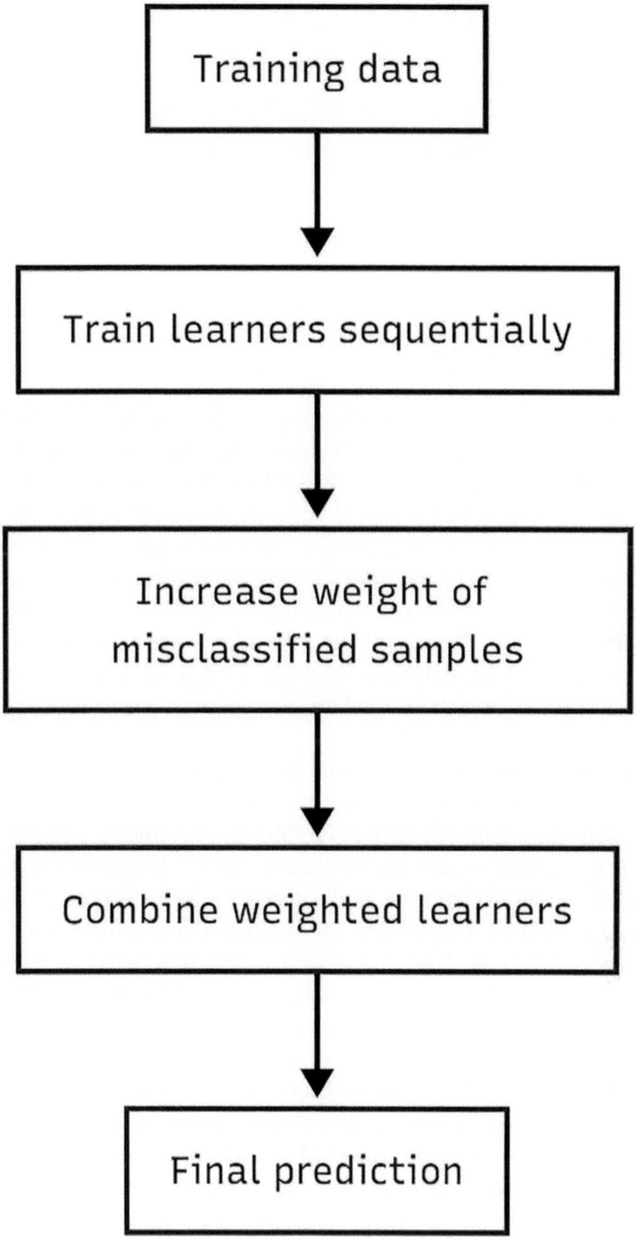


**Fig. 1.21** Boosting the ensemble principle

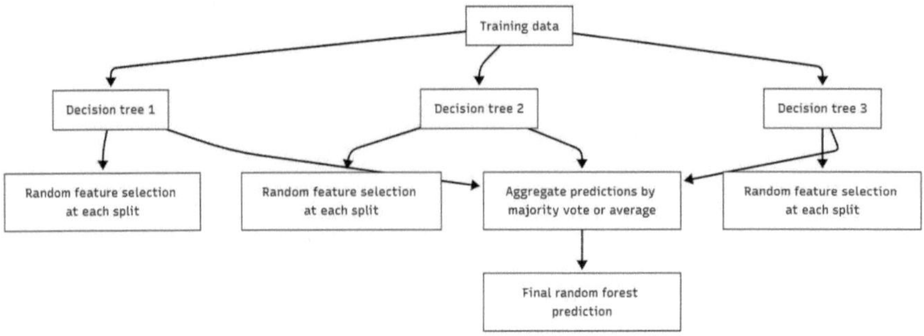


**Fig. 1.22** Principle of random forest

The parameter γ decides the system's minimum loss needed in order to divide an internal node. The larger the value of this parameter, the smaller the decision tree. In addition to this, the regularization method in XGBoost is shrinkage, and the selection of the proportion of the data to consider in each split. For more optimization in the speed of training, XGBoost incorporates a pre-ordered column-based approach for the reduction of the time spent in the determination of the optimum split by considering the attributes in parallel. XGBoost is highly optimized and simply inspects a fraction of the candidates, the percentiles of the data, and the gains are used in computing their combined statistics [80]. This makes it a very popular GB ensemble method, which is used in competitions such as Kaggle and the KDD Cup.

**LightGBM** is a comprehensive library with a pre-built histogram of features, similar to XGBoost, which utilizes gradient boosting and suggests various variants. **Gradient-based one-sided sampling (GOSS)** and **Exclusive Feature Bundling (EFB)** are the new additions in terms of features. GOSS is utilized for the construction of the base learner training sets in the ensemble. EFB is more of a preprocessing methodology used to computationally facilitate the construction of histograms for each feature at each node split. Bundling or combining sparse features in a single one allows for making it less computationally costly and feasible for GPU and CPU, as well as being fundamental and having various kinds of randomizations (column randomization, bootstrap subsampling, etc.).

LightGBM is considered faster in handling larger datasets than XGBoost, thanks to its alleviating features, while using a lower amount of memory space. In an experimental study comparing XGBoost [81], LightGBM, and other similar GBM algorithms, LightGBM with default parameters yielded the best results for different types of datasets, but its performance was uneven across datasets. By tuning the parameters further, it obtains even 75% better results than its default version. The best use cases are those involving very large datasets with high-dimensional features.

LightGBM is the fastest in general because it incorporates histogram-based learning, GOSS+EFB for faster training, but remains highly sensitive to overfitting, particularly for smaller datasets. It is also beneficial when we have limited computing resources, such as a CPU-only system, because it's lightweight and can be rapidly implemented on a fast CPU/GPU. XGBoost is a middle-ground, balanced solution because it is both tunable and stable, thanks to its well-documented nature. It provides good accuracy and tunability [82]. The good AUC performance under unbalanced datasets is particularly good when one is working in a program capable of catching irregular activity or fraud. All this still makes it highly relevant for use at competitions and practice (Table 1.4).

### *1.3.4 Hyperparameter Tuning and Model Selection*

**Hyperparameter tuning** is very important for the effectiveness of any machine learning algorithm. The effectiveness of any model largely rests upon the selection of a good set of hyperparameters, even though this is a very time-consuming and challenging task. This is done after pre-processing of the data, in an endeavor to fine-tune the sensitivity, accuracy, and robustness of the model. Comparing different settings of hyperparameters will typically be the chief way to ensure the effectiveness of the model. Two common issues in machine learning are overfitting and underfitting, which can impair generalization and accurate prediction when using new data. Overfitting is the learning of the given data too closely and performing very poorly in unseen data, and underfitting, in relative terms, takes place whenever the model is very simple and does not possess sufficient samples for appropriate hyperparameter optimization. One of the remedies for overcoming overfitting is to use trees generated at random, which tend to be less correlated [83]. However,

**Table 1.4** XGBoost or LightGBM Summary Table

| Aspect | XGBoost | LightGBM |
|---|---|---|
| Data size | Performs well on small to medium datasets; more stable when data are not extremely large | Optimized for very large datasets and high-dimensional data |
| Training speed | Slower than LightGBM due to the exact greedy algorithm and level-wise tree growth | Faster training with a histogram-based algorithm and leaf-wise tree growth |
| Memory usage | Higher memory consumption, especially with many features | More memory-efficient |
| Accuracy | Strong baseline accuracy across most tasks | Often matches or slightly exceeds XGBoost when tuned properly |
| Handling of categorical features | Requires manual encoding (one-hot, label) | Built-in support for categorical features |
| Robustness to overfitting | Level-wise growth provides more balanced trees and can reduce overfitting on small datasets | Leaf-wise growth can overfit small datasets if not regularized |
| Best use cases | Moderate-sized datasets, when stability and interpretability of tree depth are important | Very large datasets, high-dimensional sparse data, when training speed is critical |

empirical rules are mainly used to improve the robustness of models by gradually adjusting the hyperparameter, which reduces the frequency of under- and overfitting.

The first and most robust technique is called **grid search,** and it exhaustively combines all possible hyperparameter values from a specified grid. It is comprehensive and guaranteed to find the best in the given grid, but because of the exhaustive search it runs, it is computationally expensive and only works within the given grid. **Random search**, on the other hand, balances efficiency and effectiveness by not testing all possible combinations, but hyperparameters from a specified list. It is efficient and often outperforms grid search, but it might miss some optimal combinations. These two methods mostly sample without a particular guiding strategy or according to very simple rules. They can be used for a small configuration space of hyperparameters [84]. **Bayesian optimization** further constructs probability models that run through sequential trials, with each iteration upgrading the existing model with fresh data. Also, a very sample-efficient tuning method that works well for expensive evaluations, but it is very complex to integrate because it may struggle with discrete parameters [85]. **Early stopping** can also be categorized as a hyperparameter tuning technique because it prevents overfitting by stopping the process before the model starts to generalize on the validation set, thereby saving time.

**Cross-validation** is crucial for validating the model and preventing overfitting. You divide the data into different subsets and check the performance in different/ data compositions. Different procedures are employed according to the split of the data. **K-fold cross-validation** divides the data into **K groups**, which enables a balanced evaluation of the data. Stratified cross-validation for classification maintains the common class proportion in the folds, which reduces the bias in the validation. It is an extension of the k-fold validation and is very helpful in the case of imbalanced datasets. Nested cross-validation works by combining hyperparameter searching and model comparison under one framework. It allows the validation to be more trustworthy and yet less subject to overfitting.

Some of the most commonly tuned hyperparameters are the **learning rate, iteration numbers, and optimizers**. The **learning rate** is one of the most crucial hyperparameters that controls all other model parameters and how well they adjust with each iteration. Another important hyperparameter is the **max_depth**, which refers to the maximum number of levels a decision tree can have, by which the complexity of the tree can be held under control. More levels mean more complex relationships, but also possible overfitting on the training dataset. **N_estimators** is another hyperparameter that determines how many trees or estimators can be built within an ensemble model. The more trees in this case, the more accurate the Gradient Boosting algorithm will be. The downside is that it would take a longer time to train this bigger model, and there is a possibility that it will overfit our data [86]. Algorithms like XGBoost use significantly different ranges of the learning rate and n_estimators (0.001 to 0.1 and 50 to 1200, respectively). **Subsampling** also plays an important role in the process of selecting smaller subsets of data from a larger dataset to train each tree. This reduces the problem of computational intensity and time when dealing with massive datasets that are too complex for the available memory. **Colsample_bytree** is the fraction of the features (randomly selected) that are used to train each tree.

### 1.3.5 Handling Imbalanced Datasets

One would need high-quality data in order to have a working predictive model. However, real-world data typically does not improve the performance of a given model. For example, one would have class imbalance, a situation in which one sees a very high volume of instances in a given class compared to another class. The phenomenon is very common in classification-based supervised learning tasks.

Most real applications we encounter are **imbalanced datasets**. To make a diagnosis for a rare disease, for instance, datasets may consist of thousands of instances for healthy individuals and several for afflicted individuals. Faulty events are very few in fraud management as a part of fraud detection, and defects are sparse in industrial fault diagnosis, in contrast to general operating conditions. Although the minority class may appear underrepresented, it is often more significant. Although fraudulent activities and malignant cancer are statistically rare, not detecting them can cause severe damage. When we have training data that is evenly distributed over classes, machine algorithms will maximize accuracy. But when we have unbalanced data, the model will simply predict the majority class for every input and ignore the minority class, but it would still achieve great accuracy [87]. If we have a dataset that has 95% of samples belonging to one class, a naive model would achieve that level of accuracy, but it would fail to recognize the minority class. With this accuracy, the performance in these classes would be a poor measure. Instead, metrics like precision, F-1 score, and AUC would be more appropriate for this.

When we have an imbalanced dataset, we also have characteristics of it that are categorized by a significant disparity in the number of instances among different classes. Usually, we would say one class to be referred to as the majority class (dominates the dataset), while the other, the underrepresented, is known as the minority class. **Minority class** is often carrying the most meaningful or critical information. For example, within a medical dataset used for cancer diagnosis, the minority class may represent patients who actually have cancer. If a model is biased toward the **majority class** (i.e., patients without cancer), it may fail to identify critical cases, potentially resulting in delayed diagnosis and compromised treatment outcomes [88]. This challenge is exacerbated when classification models prioritize overall accuracy, thereby misrepresenting the performance on the minority class (Table 1.5).

Traditional machine learning algorithms, such as logistic regression, decision trees, neural networks, and Naive Bayes, are normally optimized while assuming class distributions are balanced. But the outcome of their performance suffers in the case of imbalanced datasets, typically by favoring the larger class and hence incorrectly classifying minority instances. This shortcoming arises due to the generalization-based learning approaches that such models use, which are not

**Table 1.5** Tuned parameters of XGBoost

| Parameter | Purpose | Typical Lower Bound | Typical Upper Bound | Notes |
|---|---|---|---|---|
| learning_rate (eta) | Shrinks the contribution of each tree | 0.01 | 0.3 | Lower values often improve performance but require more trees |
| max_depth | Maximum depth of each tree | 3 | 10 | Higher depth increases model complexity and risk of overfitting |
| n_estimators (num_boost_ round) | Number of boosting rounds | 100 | 1000 | Depends on learning_rate; more rounds needed if learning_rate is small |
| Subsample | Fraction of training data sampled for each tree | 0.5 | 1.0 | Values <1 help prevent overfitting |
| colsample_ bytree | Fraction of features sampled for each tree | 0.5 | 1.0 | Controls feature randomness |
| Gamma (min_split_loss) | Minimum loss reduction to make a further partition | 0 | 10 | Larger values make the algorithm more conservative |
| min_child_ weight | Minimum sum of instance weights in a child node | 1 | 10 | Higher values prevent overfitting by requiring larger child nodes |
| Lambda (reg_lambda) | L2 regularization on weights | 0 | 10 | Higher values increase regularization |
| Alpha (reg_alpha) | L1 regularization on weights | 0 | 10 | Can lead to sparsity in feature weights |

suitable for picking up rare but salient patterns. To address the imbalance, one should employ deliberate strategies, either by adjusting the data distribution or by modifying the learning process. Some of the major causes of data imbalance are:

1. The target class may itself occur less often.
2. Missing or incomplete data can skew the class representation.
3. In sampling for data collection, the minority class can be under-represented.
4. Minority class instances can happen near decision boundaries or extrema.
5. Volumes of redundant or unnecessary data can overpower minority patterns.
6. Noise or mislabeled instances may be substituted for minority class signals.
7. The minority cases can potentially serve as outliers.
8. Overlapping feature space between classes can blur distinctions.
9. Low-frequency small disjuncts might not be captured by the model.

To deal with the negative effect of class imbalance over the last two decades, various techniques have been introduced. All such approaches essentially fall into one of the following categories: **data-level, algorithm-level, cost-sensitive learning, and ensemble strategies** (Fig. 1.23).

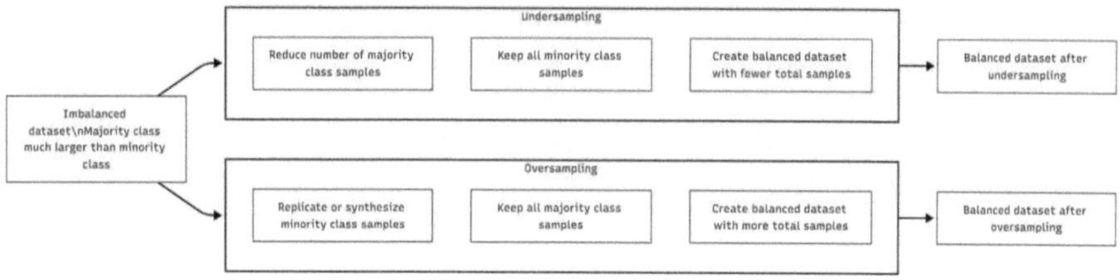


**Fig. 1.23** Undersampling and Oversampling

**Data-level methods**, also referred to as **sampling methods**, attempt to balance the set by adding more minority class samples **(oversampling)** or by decreasing the majority class samples (**undersampling**). Techniques for **oversampling** include **Random Oversampling (ROS), Cluster-Based Oversampling (CBOS), and Synthetic Minority Oversampling Technique** (**SMOTE**), which would create or repeat minority instances synthetically and balance the set. Extensions such as Borderline-SMOTE (B-SMOTE) and MWMOTE add enhancements by considering decision boundaries and hard-to-learn points. On the contrary, undersampling strategies like Random Undersampling (RUS), Cluster-Based Undersampling (CBUS), and Fast Clustering-Based Undersampling (F-CBUS) reduce the size of the majority class, thereby drawing attention to the minority class [89]. **Hybrid techniques** combine both oversampling and undersampling to benefit from the strengths of each.

**Algorithm-level methods** adapt existing algorithms in order to improve their sensitivity to the minority class. By modifying internal decision mechanisms, such as **splitting criteria** in decision trees or kernel functions in support vector machines (SVMs), these approaches can enhance the detection and classification of minority instances. Examples would include z-SVM, weighted nearest neighbor classifiers, and modified neural networks that account for class-specific error rates.

The **cost-sensitive methods** will give more weight to high-cost penalties should you misclassify minority class instances. Here, the goal is to train the classifier in a way such that it promotes minimization of costly errors, or more precisely, false negatives in high-cost areas like security or medicine. These are complementary methods alongside the algorithm-level and the data-level ones, and in most instances require the definition of a misclassification cost matrix. Some examples here are cost-sensitive SVMs, C4.5 decision trees, and PSO-based neural networks.

**Ensemble learning** was selected as a useful approach for dealing with class imbalance. These include a number of base classifiers typically trained from perturbed or resampled copies of the set, in a bid to reach a more balanced and accurate final classification. Algorithms such as **bagging, boosting, and cost-sensitive boosting** (such as AdaCost, SMOTEBoost) aggregate predictions in a way that is biased towards minority class correctness. Ensemble models are particularly effective in evening out predictions and reducing the variance inherent in small sample sizes.

Evaluating the performance of machine learning models in the context of class imbalance presents distinct challenges. Standard accuracy calculates the ratio of the correctly predicted instances to the total, but tends to be deceptive in the event that a single class overwhelms the rest. Such a model can attain high accuracy by simply guessing the majority class for all instances and completely disregarding the minority class, which typically is the class of highest concern. Thus, employing different evaluation measures is necessary in order to derive useful insights about a model's real performance in the context of imbalances. Utilizing evaluation terms like **precision, recall, and the F1-score** provides much more subtle insight into the performance of the classifier in imbalanced datasets. Precision will quantify the ratio of the positive predictions that turn out to be correct and would indicate the capability of the model in preventing false positives. Utilizing the recall, however, checks the ratio of the real positive instances that happen to be correctly characterized by the model, and is an important measure in applications where we have missing positive cases that carry serious consequences, such as medical diagnoses. The F1-score aims to be something like a harmonic mean of precision and recall, which balances both concerns and offers a single metric that accounts for both false positives and false negatives.

The Receiver Operating Characteristic (ROC) curve and the corresponding Area Under the Curve (AUC) value are useful performance measures for models trained on imbalanced datasets. The ROC curve depicts the true positive rate against the false positive rate at different threshold settings, and the AUC estimates the overall discriminability in the model for the different classes. A model whose AUC is close to 1 is highly competent, while one close to 0.5 represents performance not better than by random chance [90]. Nevertheless, for highly imbalanced data, the Precision-Recall (PR) curve and the Area Under the PR Curve (AUPRC) are more helpful in offering informative indications in many circumstances, especially in cases where the minority class is of chief interest.

There are also a few best practices that should guide the evaluation and selection of models in the context of imbalanced data. One of the indispensable steps is the use of **stratified sampling in cross-validation.** It would have the effect of ensuring that the same class distribution in the original set is maintained in each fold of the validation process, thereby avoiding performance fluctuations and providing consistency. Resampling procedures, either at the data level (such as undersampling or oversampling) or the model level (such as ensemble methods), should also be tested and compared not simply in terms of outright accuracy but in terms of the recall and the precision because they more directly suggest how the minority class is faring.

Another of the more common mistakes you should try not to make is overfitting, particularly in the case of artificial oversampling techniques like SMOTE. While such techniques enhance class balance, they may also add noise or exacerbate the effect of outliers. To counter this effect, you should consider attempting to measure models across a variety of different metrics and track performance across the training set and the validation set. Additionally, comprehension of the cost of errors in the application domain at hand can inform the choice of suitable metrics. In healthcare or fraud detection, for example, false negative minimization may be more desirable than balanced accuracy and recall.

### 1.3.6 Concept and Foundations of Transfer Learning

**Transfer learning** represents a significant advancement in machine learning, which is characterized by its ability to transfer knowledge from a well-learned source task to a less-resourced target task. When comparing conventional approaches and transfer learning, transfer learning minimizes the need for extensive target data by leveraging patterns, representations, or parameters learned in related source domains, while conventional approaches require large amounts of labeled data and independent training for each task.

The key lies in identifying and transferring knowledge of common task characteristics. A general object classification in an image-trained model can, for instance, be fine-tuned for targeting specific medical imaging tasks and hence significantly reduce the computation and data gathering overhead. The framework involves pretraining the network on a source task for which abundant labeled information is readily available, followed by fine-tuning the network to achieve optimal performance on the target task, for which limited data are available [61]. The framework is particularly useful in the context of supervised learning applications in which target-labeled examples in general are scarce or expensive.

The core of transfer learning lies in **aligning the domain and task**. The domains here point towards the label and feature spaces, while the tasks include the target functions or the predictions. Transfer learning is effective in the case of having similarity across the source and target domains, such that the transferred knowledge is useful and relevant. However, effective mechanisms should also counter domain differences and avoid adverse transfer, in which knowledge transfer at the wrong level can adversely affect target task performance. This framework has also been widely applied in various fields, including natural language processing, computer vision, and biomedical research, where the issue of insufficient data is a prevalent concern. Transfer learning continues to expand the frontier of possibilities in machine learning in resource-limited applications (Fig. 1.24).

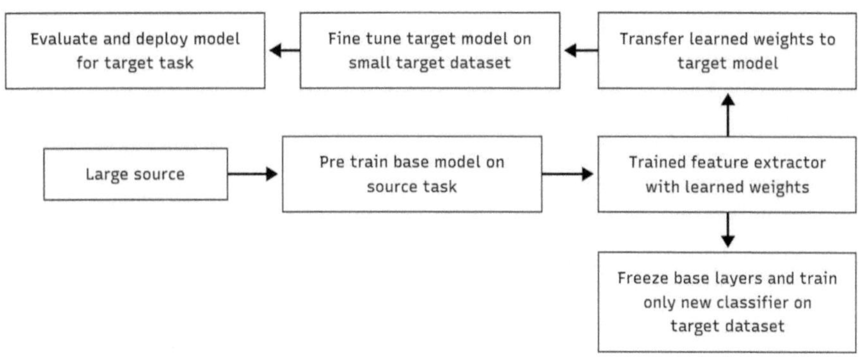


**Fig. 1.24** Block diagram of transfer learning

One of the basic techniques in transfer learning is **instance-based transfer**, where emphasis is placed on reusing specific data instances from the source domain to improve the performance in the target domain. Using this method, we are positing certain instances are related to the target domain, and we are assigning weights to them in a way so as to match their distribution in the target domain [62]. Methods such as **instance weighting, sample selection bias correction, and reweighting frameworks** (e.g., TrAdaBoost) are popularly used. These procedures work best when the source and target domains have similar space features but have differing marginal distributions.

**Feature-based approaches** focus on identifying and transforming features from the source domain to match the target domain. These transformations may involve feature augmentation, dimensionality reduction, or mapping to a latent space. Methods such as **Transfer Component Analysis (TCA), Joint Distribution Adaptation (JDA), and autoencoder-based techniques** create shared feature spaces to facilitate this knowledge transfer. Feature-based methods are especially useful when the feature spaces are similar, but distributions or label relationships differ between domains.

**Parameter-based approaches** will transfer knowledge by reusing the parameters or structure of pre-trained models. This category is more inclusive, fine-tuning pre-trained neural networks and multi-task learning frameworks where shared parameters are optimized for multiple related tasks, essentially focusing on how to incorporate the pre-existing model. These methods are highly effective in scenarios where pre-trained models encapsulate generalized knowledge applicable to the target domain.

**Relational-based transfer** methodology centers around transferring logical relationships or logical rules across the source and target domains. These methodologies are particularly effective in domains where the internal structure, such as graphs or relational databases, provides valuable insights for learning [63]. Relational-based feature mapping and knowledge graph embedding are a few such examples.

A paradigm shift in the field of machine learning is Transfer learning, as it offers innovative solutions to challenges such as limited labeled data and high computational costs. By leveraging knowledge from pre-trained models or related tasks in our research, transfer learning enables efficient learning, a valuable tool in data-scarce environments. However, despite its advantages, it also presents significant challenges. By understanding these advantages and challenges, we can find effective solutions and applications [64].

Pros of Transfer learning are:

- Reducing training data: Since transfer learning utilizes existing knowledge, this would enable the model to perform better even when we possess limited data. For instance, a pre-trained language model, such as GPT, fine-tuned from a small dataset, can perform better than newly trained models.

- Quicker training times: Since we already possess some fundamentals in the way of knowledge in the model, training the task would involve fewer iterations and less labor than constructing the model from scratch.
- Better generalization: It will mix representations and features across tasks, and this will increase the model's capacity for generalizing unseen target domain data.
- Cost-effectiveness: Transfer learning reduces the need for extensive data collection, annotation, and computational resources, making it a cost-efficient choice for many real-world applications.
- Broad applicability: Transfer learning exhibits universality in many application domains like computer vision, natural language processing, and healthcare, and facilitates acquiring breakthroughs in applications from image classification to disease diagnosis.

Limitations of Transfer learning:

- Negative transfer risk: We need source and target domains or tasks to highly overlap, or they can lead to negative transfer, as this transferred knowledge can degrade instead of enhancing performance.
- Computational overhead: Transfer learning is computationally costly and requires a great deal of memory.
- Pre-trained model dependence: Transfer learning is very much dependent upon access to high-quality pre-trained models for many domains or tasks for which they may not exist.
- Fine-tuning and overfitting the target data: In certain scenarios, fine-tuning can lead to overfitting the target small target set and thereby defeat the purpose of transfer learning.
- Interpretability challenges: Increasingly more complicated models make it more difficult to discern the workings of knowledge transfer and draw inferences about model activity.

## 1.4 Unsupervised Learning Techniques

**Unsupervised learning** is a machine learning subfield that addresses finding patterns in data without using labeled examples. Unlike other processes that demand direct instructions or pre-defined categories, unsupervised learning examines the data itself to identify groups, similarities, or underlying patterns. Its core function is to find the relations or patterns that are not necessarily easily visible, and in doing this, valuable insights are revealed. This is especially helpful when dealing with a massive amount of data. By organizing and summarizing the data, unsupervised learning allows researchers and analysts to better understand the most significant features in the data. This typically leads to the identification of new findings and provides support for making more informed decisions [91, 92]. This occurs even when nothing at all about the data is known. Some of the most common techniques applied in the application of unsupervised learning include [93]:

- Clustering—used for bundling the points having similar attributes in order to discover different groups in the data.
- Dimensionality reduction—helps compress high-complexity datasets by reducing the number of variables involved, making them more manageable. Such algorithms as PCA, t-SNE, and UMAP are useful in compressing the number of variables in a dataset without deleting its useful characteristics. In this context, such algorithms facilitate the interpretation and visual identification of patterns that cannot be observed directly.
- Anomaly detection—applied to identify out-of-the-ordinary or unusual data points that do not fit common trends, and can prove highly efficient in use cases such as fraud detection.

Unsupervised learning algorithms also appear in many applications. Finance applies them in customer clustering for appropriate categories and medicine in identifying patient groups whose states of illness are comparable. Unsupervised learning is a valuable approach, although it presents a few challenges. One significant challenge is the estimation of the quality of the results, as pre-classified labels or categories cannot be cross-verified for corresponding validation in an a priori manner [94]. What this implies is the need for clear knowledge and context in the interpretation of the outcomes. Unsupervised learning is an effective technique for exploration of the data and the extraction of new knowledge, and it can act as an input for better decision-making in many areas.

### *1.4.1 Clustering Methods*

**K-Means Clustering** is a type of unsupervised machine learning technique used to divide unlabeled data into a small number of groupings. It is used in classifying data according to how similar they are. The algorithm first selects a few centroids or center points at random to form a cluster. All the points are then classified according to their nearest centroid, based on how similar they are. The centroids are updated by taking the average locations of the points in the clusters after a point is assigned. The same continues until the centroids cease to move and create clusters [95]. Dividing the points in the data into clusters such a way similar points are in the same category is the purpose of clustering.

To understand how it functions, we must consider it in a practical context. A vector-like set of objects with certain characteristics and respective values for those characteristics is presented to us. Clustering those objects in groups is the task. The algorithm employed here for this task is the K-means algorithm. The value of groups or clusters we would like to split our stuff into is reflected in the term "K" in the algorithm's name. The objects will then be partitioned into k groups or similarity clusters by the algorithm. We shall employ the **Euclidean distance** for the purpose of such similarity. The algorithm then operates in the following way:

- We begin by assigning the points k initially referred to as the cluster centroids or the means at random.
- Each point is grouped by the nearest mean, and the coordinates of the mean—that is, the averages of the objects already assigned in the given cluster—are updated.
- Once a specified number of iterations have taken place, we repeat the procedure until we reach our clusters.

One such method by which you can achieve the perfect number of clusters is the **elbow approach**. An elbow method, a graphical method for ascertaining the perfect number of clusters for a k-means clustering algorithm, is the elbow approach. You calculate the euclidean distance of each point in the data set from the cluster center and the number of clusters by finding the point at which the increase in the **"within cluster sum of squares" (WCSS) plateaus**. The result is obtained by plotting the total variation within each of the clusters against the total clusters. The initial step in the elbow method is the identification of the WCSS for each cluster (k). The second step is the presentation of the number of clusters along the x-axis and the WCSS value along the y-axis. The points in the plot should have a smooth nature as the value increases for the clusters. The optimal value of the cluster can be obtained by this technique [96, 97]. Keep the computational cost in mind while choosing the value of the clusters. The power of processing increases with the addition of the value of the clusters, particularly while processing the very huge datasets (Figs. 1.25 and 1.26).

Each clustering method that shall be mentioned has its own uses and proves more effective than others in specific situations based on our needs. Almost every sector

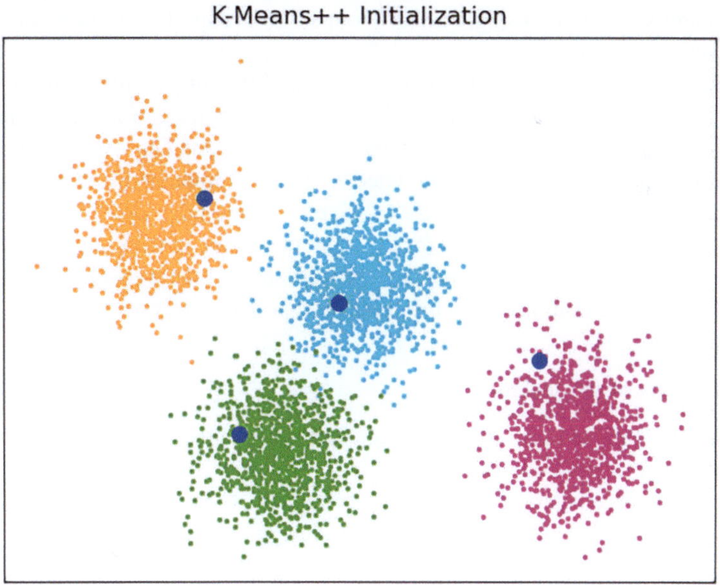


**Fig. 1.25** K-means clustering (https://commons.wikimedia.org/wiki/File:K-means%2B%2B.png)

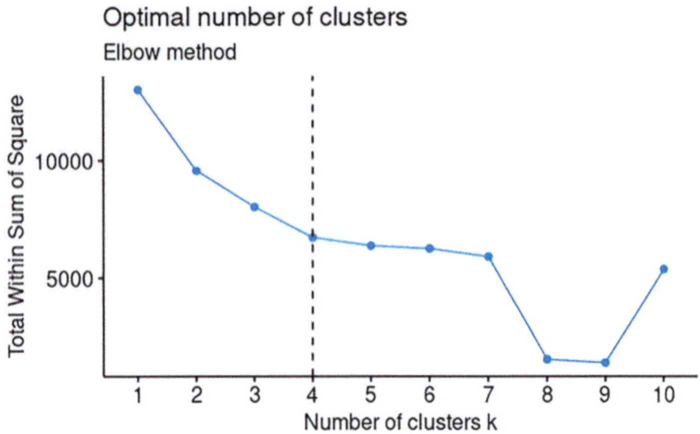


**Fig. 1.26** The elbow method (https://commons.wikimedia.org/wiki/File:Elbow_method.png)

and subject uses the **k-means clustering technique**. It is usually used for machine learning data that is readily portionable, numerical, and has a few dimensions [98]. Researchers have integrated k-means clustering and deep networks, such as CNNs and RNNs, to enhance performance for a highly diverse set of tasks in machine learning, including computer vision, natural language processing, and numerous other applications. Some real-world applications of K-means clusters are:

- Customer segmentation—the act of dividing a business's customers based on common characteristics that suggest similarity. These companies can employ this method to target a single set of groups or a group of customers for a single marketing campaign.
- Document classification—the act of assigning different classifications or categories to documents. It is a method used by many businesses for moderating their content. To create a document classifier, try this Watson Discover tutorial.
- Computer vision method—abbreviated as image segmentation, segments a digital image into individual clusters of pixels. Image edge detection in medical images is explored in this work for the implementation of the k-means algorithm.
- Recommendation engines—used by software around the web. Online stores utilize k-means clustering and principal component analysis to make product suggestions.

A few benefits of this technique are:

- Simple to learn and use: It is a very uncomplicated algorithm. It is the best algorithm for unsupervised machine learning.
- Fast: K-means clustering is developed through an iterative, computationally simple approach. The algorithm for k-means clustering is faster than for hierarchical clustering, which involves the computation of the pairwise distance between each datapoint and the development of a tree-like structure of groups.

- Scalable: K-means is scalable for clusters of any shape and size, can easily be scaled for very large datasets, and is also appropriate for cluster research. The algorithm is more scalable and suitable for very large datasets compared to the previous methodologies, owing to high computing efficiency.

Whereas the cons are, for instance [99, 100]:

- Parameter dependency for input: K-means clustering uses properly defined input parameters. To attain desirable cluster outcomes, the suitable centroid and the value for the clusters should be initialized. Incompatible initialization of the centroid might result in inferior cluster assignments and longer run times. Many studies have been directed at improving the method of initializing the centroids in a bid to attain faster convergence times and better clusters.
- Potentially poor performance on specific datasets: When there are similar-sized clusters in the dataset and no significant outliers or density fluctuations, K-means works well. When the dataset is highly dimensional or has a lot of variance, K-means performs badly. K-means may generate low-quality clusters if the data does not match specific dataset assumptions. Unevenly sized clusters, for instance, may cause bias and incorrect categorization among the smaller clusters by tilting the centroids toward the bigger clusters. This issue can be resolved by generalizing k-means with probabilistic models such as the Gaussian Mixture node.
- Significant outlier impact: The outcomes of k-means clustering are significantly impacted by outliers. Although they shouldn't be too widely apart, different clusters shouldn't distort the results. Before using k-means, it is crucial to take the data assumptions into account. Since K-means seeks to identify centroids by averaging data within a cluster, it is particularly susceptible to outliers. Because of its sensitivity, it is likely to overfit in order to incorporate these outliers.

**Hierarchical learning** is an unsupervised type of machine learning algorithm that partitions the data into a tree of nested clusters. The two prominent forms are agglomerative and divisive.

- Dividing top-down or dichotomously in such a way that the entire set is repeatedly bisected in a tree-like manner.
- Divisible or top-down method 2 starts by placing the complete set of information in one cluster and then continues by dividing it into smaller and smaller clusters till the cluster contains a singleton.

**Hierarchical cluster analysis** helps identify connections and patterns within datasets. The results are presented in the form of a dendrogram diagram showing the distance relationships between clusters. It operates by placing objects in a hierarchy of clusters without a strict hierarchy. Hierarchical procedures for clustering are applied in various disciplines, including the social sciences, biology, and image analysis, to study and identify patterns in the given information. Prospective applications include identifying communities within network models of a specific node type, segmenting customers, and classifying clinical research populations [101,

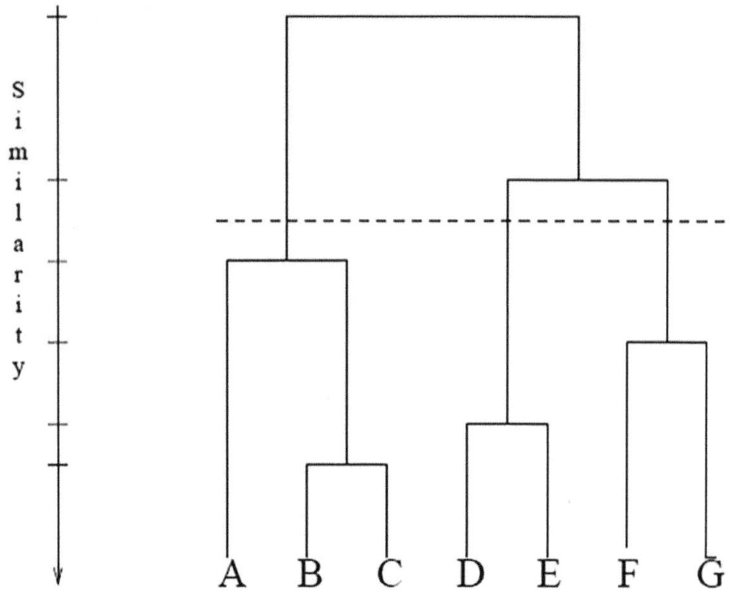


**Fig. 1.27** Dendogram (https://commons.wikimedia.org/wiki/File:Agglomerative_clustering_dendogram.png)

102]. The **dendrogram** is a tree-like figure often used to represent the hierarchy of groups. It shows the similarity or the distance among the points of the data and the progression in the splitting or combination of the clusters. Dendrograms may also refer to a nested list of lists by attributes. The y-axis or line height of the dendrogram shows how far apart the clusters were at the point of combination, while the x-axis presents the points of the data [103] (Fig. 1.27).

To determine the number of clusters in your final clustering model, you may utilize a dendrogram. Finding the tree's natural cutoff point, when the branches naturally shrink and lengthen, is one tactic. Alternatively, when a horizontal line cuts the dendrogram, the number of vertical lines crossed indicates the number of clusters.

Some possible applications for this kind of clustering are the following [104]:

- Business—Through the categorization of the customer details along product preference, demographics, spending behavior, risk profile, or social activities, for instance, hierarchical or agglomerative clustering facilitates the examination of trends and divisions of the data.
- Bioinformatics and clinical research—Patient groups for clinical research may run in the thousands. Hierarchical clustering utilizes variables such as DNA mutations, physiological responses, or clinical criteria to categorize different populations into more homogeneous groups. To understand evolutionary advancements, it can also be employed in organizing species by biological attributes.

- Image and information processing—In text recognition applications based on images, hierarchical clusters are employed for classification of shape-based groups of handwritten letters. It can also be utilized for classifying search outcomes or for the storage and recovery of information in terms of specified terms.

**Density-Based Spatial Clustering of Applications with Noise (DBSCAN)** is a density-based clustering method that classifies outliers as noise according to their density in the feature space and clusters data points that are densely packed together [105]. Clusters are defined as dense areas in the data space that are divided by less dense areas. In contrast to K-Means or hierarchical clustering, which presume that clusters are spherical and compact. DBSCAN works ideally for real-world outliers in real-world data, such as:

- Clusters may be random in shape; they need not be convex or spheroid.
- Noise and Outliers: It does not impose any cluster upon the noise points and hence detects and processes them.

To grasp DBSCAN, we need to understand two very significant parameters [106]:

- eps: It determines the radius of the neighborhood around a point in the data. Two points are then seen as neighbors the moment their distances are smaller than or equal to the value of eps. Looking at the k-distance graph is a popular method for determining the value of eps. Finding the right value for eps is important:—MOST points end up as noise once the value of the eps is too small.—Clusters can run together once the value of the eps gets too great and the algorithm cannot distinguish them.
- MinPts: It is the absolute minimum of points for the formation of a dense zone within the eps radius. It is a common practice to set the MinPts > = D + 1, where the value of D represents the number of dimensions in the dataset. Thus, for the majority of the scenarios, the MinPts would be 3 because the highest common value for the number of dimensions is 2.

How DBSCAN works is by clustering the data into three groups.

1. Core points having at least a requisite number of neighbors within a specified epsilon radius
2. Border points near core points, but having too few neighbors for them to be core points
3. Noise, which are points that are not part of any cluster (Fig. 1.28).

Once the significant parameters are understood, it's pretty simple to, in fact, build DBSCAN clusters [107, 108]. The standard procedures are:

1. Core point identification—Count the points in the eps neighborhood of any point in the database. The point is a core point if the count is greater than or equal to the value of MinPts.
2. Clustering—Form a new cluster for a yet unassigned core point. Add to the cluster all the points which are density-reachable from the core point—that is, points in the eps radius of the core point—recursively.

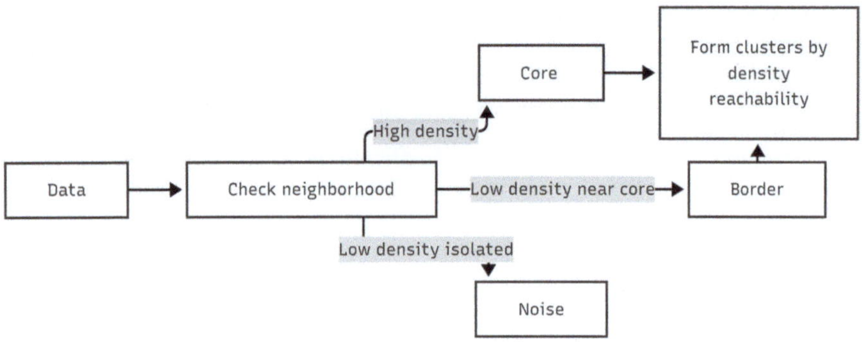


**Fig. 1.28** The three clusters of data in DBSCAN

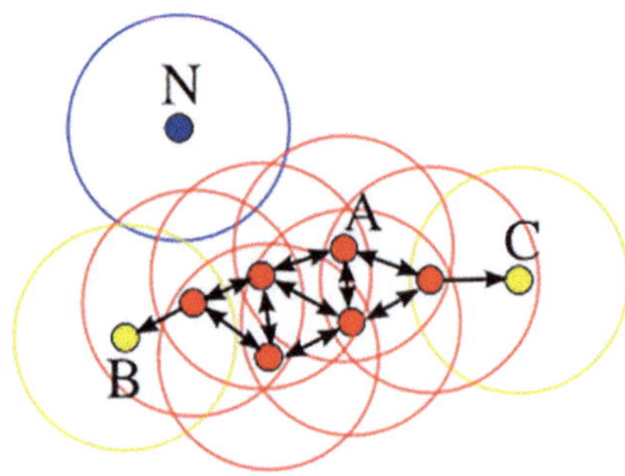


**Fig. 1.29** Clustering by DBSCAN (https://commons.wikimedia.org/wiki/File:DBSCAN-Illustration.svg)

3. Density connectivity—If a chain of points exists such that at least one core point and each point of the chain is in the eps radius of the following point, then any two points a and b are defined as density-connected. Continuing a series of dense areas, this chaining method ensures each point in a cluster remains connected.
4. Noise point labeling—A Point not belonging to any cluster is annotated as noise during the final processing of points (Fig. 1.29).

DBSCAN offers several benefits, particularly when working with **complex and noisy datasets**. Its resilience to outliers is one of its main advantages, as it reduces the impact of anomalies by treating single points as noise and constructing clusters based on dense data regions. In contrast to many conventional clustering techniques, DBSCAN is more flexible and adaptable to a variety of datasets since it does not need the number of clusters to be predetermined. It is also highly effective in identifying groups of arbitrary forms, which makes it a good fit for data with non-linear or irregular structures [109]. The method also has the advantage of having only two primary hyperparameters: min_samples (the minimum number of points required to build a dense cluster) and eps (the maximum distance for neighborhood consideration), which facilitates practical implementation and makes tuning easier.

Despite its benefits, DBSCAN also possesses several disadvantages that can hinder its effectiveness and performance. One major concern is sensitivity towards parameter selection of hyperparameters, specifically the distance threshold (eps) and the number of points (min_samples) required for creating a dense region. Wrong parameter settings can cause under-segmentation, i.e., merging of different clusters together, or cause a single cluster to be split into a few sub-clusters. Additionally, clusters whose densities vary create challenges for DBSCAN. Inaccurate or disjointed cluster formation may be the result of not being able to match sparsely populated areas with the appropriate clusters. Failing to make accurate predictions or generalize and apply effectively to new and unseen data is also a disadvantage. It does not facilitate the use of prediction, unlike some other forms of clustering. What this implies is the inability, after having already clustered a collection of items, for it to also give cluster membership to more points in the collection without having to fully retrain the model.

There are numerous uses of DBSCAN, particularly in scenarios in which controlling the noise and detecting clusters with arbitrary shapes are important. Because it can detect irregularly shaped clusters, it is very effective in the analysis of spatial data. It is therefore effective in geographic information systems (GIS) for operations such as aggregating places for activity patterns or locating areas with similar land use. Due to the inbuilt capability of differentiating the core points and the outliers, it also finds application in the identification of outliers or the identification of points of anomalies. It is employed in customer segmentation in the fields of business analytics and marketing, allowing for the differentiation of consumer groups with similar preferences or consumption behaviors [110]. It finds application in tracking and aggregating areas with similar levels or characteristics of pollution or environmental conditions for environmental studies. It facilitates the identification of the hotspots for congestion and the classification of the routes with similar traffic flow in transport and traffic studies.

**Dimensionality reduction** is a significant component of the unsupervised learning method. It is defined as the act of limiting the number of features or variables in a set of data while maintaining as much useful information as is feasible. **High-dimensional data** is difficult to understand and visualize, making it challenging for analysis. Dimensionality reduction, therefore, facilitates ease of understanding and utilizing the data [111] (Table 1.6).

There are various dimension reduction techniques, and the top ones in use are **PCA (Principal Component Analysis), t-SNE (t-distributed Stochastic Neighbor Embedding), and UMAP (Uniform Manifold Approximation and Projection).** These are different in how they work and in the context in which they are helpful. **PCA** is a linear method whose main purpose is to reduce the data by projecting it onto a new set of axes known as principal components. These components are the axes in the directions in which the greatest variations in the data occur [112]. PCA captures the global structure of the data and is very speedy and easy to interpret. It can only discover linear patterns and, therefore, may not work effectively for more complicated, non-linear data.

**t-SNE** is a non-linear method that pays special attention to the conservation of the local connections between points. It operates by transforming distances between points into probabilities and then attempting to preserve similar points in the reduced

**Table 1.6** Comparison of Clustering method

| Method | Key Idea | Input Requirements | Strengths | Limitations | Typical Use Cases |
|---|---|---|---|---|---|
| K-means | Partitions data into $k$ clusters by minimizing within-cluster variance | Must specify the number of clusters $k$ | Simple, fast, and works well on spherical clusters | Sensitive to initialization and outliers, assumes equal variance | Customer segmentation, document clustering |
| Hierarchical (agglomerative/ divisive) | Builds a hierarchy of clusters using distance metrics | Distance matrix or raw data | No need to predefine cluster count, dendrogram gives flexible cuts | Computationally expensive for large datasets, the choice of linkage affects the results | Gene expression analysis, taxonomy |
| DBSCAN | Groups of points with high density separate the noise | Requires $\varepsilon$ (neighborhood size) and MinPts | Finds arbitrary-shaped clusters, identifies noise | Parameter tuning is critical, struggles with varying densities | Spatial data, anomaly detection |
| Gaussian mixture model (GMM) | Fits data as a mixture of Gaussian distributions using the EM algorithm | Number of components | Soft clustering (probabilistic) can capture ellipsoidal clusters | Sensitive to initialization, assumes a Gaussian distribution | Image segmentation, speaker recognition |
| Mean shift | Moves cluster centers toward regions of high density | Bandwidth parameter | Automatically finds the number of clusters, handles non-spherical shapes | High computational cost for large data | Computer vision, mode detection |

space near one another [113]. t-SNE is excellent for visualizing clusters and unsuspected structures in the data at a low or very low dimensionality.

**UMAP** is a newer algorithm of a non-linear kind. It attempts to preserve both the global structure and the local structure in the input data, and is faster and more scalable than t-SNE. UMAP is based on mathematical theory in topology and in the field of geometry. UMAP works very well for larger datasets and produces better and more stable results than t-SNE; however, it can be sensitive to parameter settings. Both have pros and cons. PCA is very fast and easy, but restricted to linear relationships. t-SNE is very good for visualization, but slow and not very stable. UMAP balances speed, accuracy, and scalability very well [114, 115]. The method to use depends on the objective of the analysis: for straightforward reduction or compression of the characteristics, PCA suffices in general. To visualize clusters and high-level structures, t-SNE or UMAP is recommended [116].

**Anomaly detection,** also referred to as **outlier detection**, is an indispensable method employed in the identification of points in any given set of data that are unusual from common patterns. In most cases, genuine anomalies are rare; hence, this process is typically done without pre-labeling how the anomaly appears. Anomaly detection is hugely significant in many fields, such as fraud identification and network security, where the mere identification of anomalous activity or unusual patterns can indicate new issues or threats, thereby safeguarding systems and data [117]. Anomalous detection also encompasses numerous distinct techniques for the identification of unusual points in the data. One such popular method is a Distance-based one; i.e., the method estimates how far the points in the data are from the supposed average or common data (Fig. 1.30).

The **Mahalanobis distance** computes the distance from the centre of the normal distribution of the data, considering the correlations between the variables. If a point finds itself across one such barrier, then it is indicated as an anomaly. The second collection of techniques pertains to density. The Mahalanobis distance D from a point x to a distribution mean μ having a covariance matrix S is defined by the following formula:

$$D = \sqrt{(x-\mu)^T S^{-1} (x-\mu)} \tag{1.20}$$

Here, x is the vector of the point being measured, μ is the mean vector of the distribution, and S-1 is the inverse of the covariance matrix. This formula helps scale the distance based on the distribution of the data and is a useful measure in anomaly detection because it indicates how far a data point is from the average value of a group, while also considering the overall shape and spread of the data. Unlike simple distance measures, it takes into account the relationships between variables, so points that lie far from the center according to this distance can be identified as possible anomalies [118]. One of the key advantages of Mahalanobis distance is that it is not affected by the scale of the variables, meaning it works well even when the data includes measurements with different units or ranges. It also handles situations

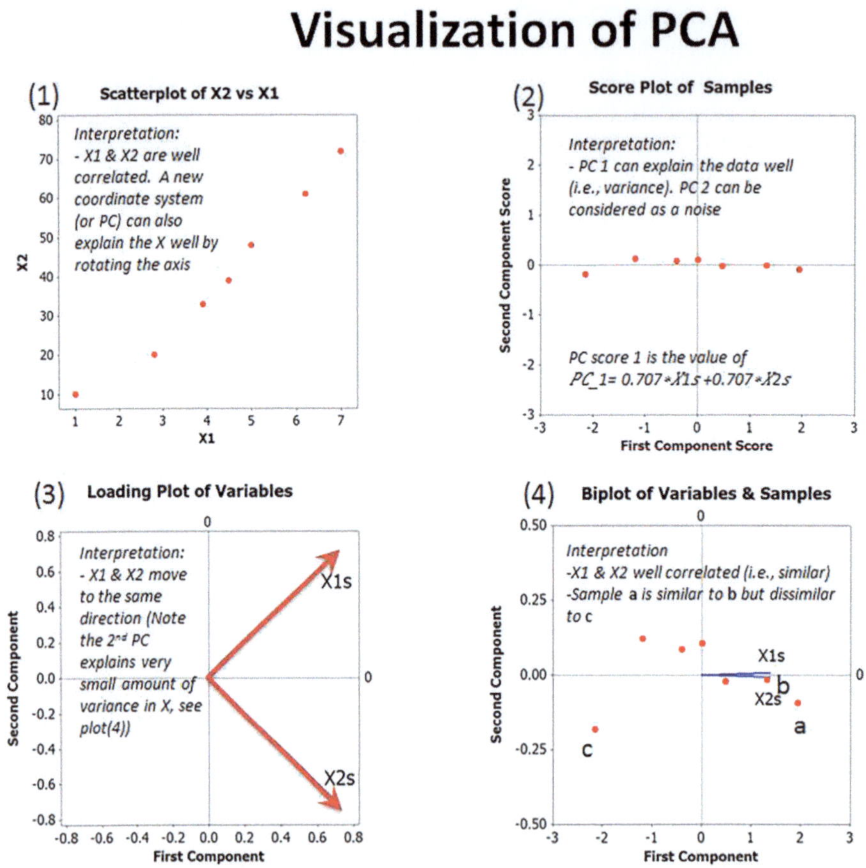


**Fig. 1.30** PCA Visualization (https://commons.wikimedia.org/wiki/File:Simple_Example_of_Visualization_for_PCA.jpg)

where variables are correlated, which often occurs in real-world datasets, making it a reliable tool for identifying unusual data points in complex datasets.

These density-based techniques, such as the **Local Outlier Factor**, look at how crowded or sparse the area around a data point is. Points that lie in regions with fewer neighbors than expected are considered outliers because they don't fit well with the surrounding data. One also possesses the model-based techniques for which a model of the typical look of the data is built. Such algorithms as Isolation Forests and one-class Support Vector Machines (SVMs) build models of common activity, then classify new points in the data to determine whether they are unusual. These techniques work best in the case of unusual points occurring very infrequently and not being clearly defined in advance.

Each of these methods excels in its own right and is appropriate for various forms of data and tasks. Distance-based approaches are good for cases in which the distribution of the data is relatively well known, the best approaches for point-wise deviations are the density-based ones, and the model-based ones give flexible means of accommodating difficult normal patterns.

The Mahalanobis distance is a useful statistical metric that finds extensive use in various contexts, particularly for identifying outliers in multivariate data. Since the Mahalanobis distance also considers correlations in the variables, it is a more trustworthy metric than the more straightforward Euclidean distance in the context of datasets having many related features. In real-world image processing applications, a primary task is the differentiation of objects of interest from noisy, complex backgrounds. Classification and segmentation of image areas based on color, texture, or intensity is a ubiquitous task in various fields. The application of the Mahalanobis distance method is common in such tasks because it can estimate statistical similarity considering the correlation of the characteristics. It operates effectively in scenarios involving object detection, background removal, and anomaly segmentation, both in natural and synthetic environments.

For example, the extraction of a specific type of vegetation from neighboring soil or a tumor in a medical image requires something more than simple thresholding. Proximity measures such as the Euclidean distance weigh all the attributes equally and ignore the reality in world data that usually consists of attributes that are interdependent or have different degrees of variance. The Mahalanobis distance adjusts for the statistical structure of the data by considering the variance and covariance of the input attributes and, therefore, sees widespread use in classification applications in which accuracy and robustness are more important.

The application of Mahalanobis distance in image classification is typically a multi-stage process that involves data representation, statistical modeling, distance computation, and decision-making based on these distances:

- Color feature vector representation: Every pixel in a color image is represented by a feature vector. In RGB images, this kind of feature vector contains three values:

$$x = \begin{bmatrix} R \\ G \\ B \end{bmatrix} \tag{1.21}$$

- You can use these RGB levels straight away or convert them to any color space like HSV, LAB, or even the YCbCr, as the nature of the segmentation task might dictate.
- Choice of the training sample: A collection of training pixels for which the membership in the current object under examination is specified manually or automatically. The training pixels serve for estimating the statistical parameters in the specific Mean vector μ and Covariance matrix S:

$$\mu = \frac{1}{n}\sum_{i=1}^{n} x_i ; S = \frac{1}{n-1}\sum_{i=1}^{n}(x_i - \mu)(x_i - \mu)^T \tag{1.22}$$

- Here, $x_i$ is the RGB value of the i-th pixel in the training samples, and n is the total number of training samples.
- Calculation of Mahalanobis distance: The Mahalanobis distance of a pixel value x across the whole image to the vector of mean color μ is given by this formula:

$$D_M(x) = \sqrt{(x-\mu)^T S^{-1}(x-\mu)} \tag{1.23}$$

- This equation calculates how far a pixel's color is from the target object's average color, considering the shape and orientation of the data (i.e., warping/compressing distances based on feature interdependencies).
- Classification decision: The Mahalanobis distance of a pixel is then computed, and a threshold T is used. When DM(x) < T, the pixel is classified as the object; otherwise, it is classified as the background. The two-class decision rule produces a segmentation mask that delineates the areas in the image statistically closest to the training class.
- Noise elimination and post-processing: For enhancing the quality of the segmentation result, one may use adaptive filtering or morphological operations (e.g., erosion, dilation, or connected-component labeling). These filters eliminate the isolated misclassification pixels (noise) and close the small holes in the segmented shape, yielding cleaner and more precise output.
- Actual use: The algorithm for the Mahalanobis distance was implemented in a few research and commercial software packages. In one application for analyzing agricultural images, a program having a GUI was written, which performed this task on the images of rice crops in paddy fields. The user manually selected the training areas in the image, and then the program calculated the desired statistics and performed pixel-wise classification.

Results obtained under such mappings have demonstrated high accuracy rates in delineating green crop regions from brown soil or water regions, even under different illumination settings. Such insensitivity originates from the fact that the algorithm can be trained to adapt to the color variation of the crop pixels and distinguish them from background noise [119]. For the same applications in clinical imaging employed for the delineation of tumors or lesions in an MRI or CT based upon differences in intensity or texture, since pathologic areas characteristically differ slightly in a healthy tissue, the application of Mahalanobis distance in the characterization of these differences in terms of a fraction of a healthy training set enables a more sensitive identification than the common approaches.

A major challenge in the use of the Mahalanobis distance is the assumption that the data set originates from the multivariate normal distribution. In practice, however, for the majority of applications, this assumption does not hold exactly. This is undesirable because the outcomes of the Mahalanobis may not be safe for use with non-Gaussian data. The data may then be transformed in some manner in an attempt at a better approximation of the normal, or a different measure of distance less affected by non-normality may be used. Lastly, the Mahalanobis distance is a highly powerful and flexible tool used for identifying anomalies in a wide range of fields. The good thing about the Mahalanobis distance is the point that the correlation of the variables is also considered. It does possess one low point, however: extensive testing is needed in order for such a methodological assumption to be revealed.

**Isolation Forest** is a powerful unsupervised learning method for anomaly detection, leveraging the fact that anomalies are less common and more different, and therefore easier to find via random partitioning. The algorithm constructs multiple isolation trees, and each tree is constructed by recursively selecting random features and split points. Anomalies, owing to their unique nature, tend to be singletons with fewer splits and thus have shorter path lengths between the root and leaf nodes [79]. This method not only provides efficiency but also excels in terms of noise reduction and eliminating irrelevant features. The algorithm produces several decision trees or isolation trees. The tree is constructed by choosing a random attribute in the set and a random split value in the interval between the minimum and maximum value of the attribute chosen. The algorithm is repeated and applied several times until the entire set of points is split or the desired tree depth is obtained. The reasoning behind this approach is that the anomalies will normally need smaller splits in order to isolate them from the normal points [75]. The result is that the length of the path from the tree's root node to the leaf node at which a point gets isolated is smaller for the anomalies. It is a very important measure used in classifying whether or not a point is an anomaly.

The Isolation Forest is employed in cybersecurity applications, where it recognizes unusual activity in network traffic, indicating potential intrusion attempts or cyberattacks. The data in such systems tends to arrive in the form of logs or streaming events, and the algorithm's capacity to successfully detect unusual patterns, such as unusual login hours, access to sensitive records, or high volumes of data transfers, is essential in ensuring security. The algorithm's versatility in responding to dynamic environments without requiring labeled training data makes it a potent tool in the field. Isolation Forest consists of the creation of many random trees for isolating the points. Normal points must have many breaks for them to isolate them, but a few partitions quickly isolate the anomalies. Such a phenomenon leads to boundaries that isolate anomalous and normal areas.

Manufacturing and mechanical engineering also apply Isolation Forest for monitoring equipment and processes. Industrial equipment generates a wide variety of sensor readings, indicating its normal operation. With constant monitoring, Isolation Forest can spot the early signs of failure or deterioration in operations. For example, a minor fluctuation in the vibration patterns or temperature readings can turn out to be a sign of major wear or imminent failure [79]. This allows maintenance teams to take action similarly in anticipation of bigger breakdowns, saving time and the associated cost of unnecessary downtime.

Isolation Forest has also found usage in defense and intelligence for surveillance data analysis. Tracking satellite data, radar returns, or communications feeds, the algorithm is used to identify activities that deviate from regular behavior. For instance, an unusual change in vehicle activity within an otherwise quiet area or an unusual communications spurt could be an indication of a threat. The algorithm's ability to identify rare events makes it a valuable tool for enhancing situational awareness and decision-making in high-stakes situations.

**Association rule learning** is a significant algorithm in the field of unsupervised machine learning, used for identifying relationships and patterns in large datasets. The application of this algorithm is quite common in market basket analysis, wherein the algorithm is utilized for the identification of the kind of products purchased together by customers. For instance, a customer who purchases bread and butter might also purchase jam. In this way, identifying such patterns helps enhance marketing campaigns, recommendation systems, and inventory management in businesses (Table 1.7).

In order to determine the strength of association rules, **support, confidence,** and **lift** are employed. **Support** indicates the frequency the item or itemset appears in the database. The **confidence** reveals how many times the rule was discovered to happen, and **lift** reveals how much more probable the purchase of the second item becomes upon the purchase of the first item than by random chance. The two algorithms for association rule learning are Apriori and FP-Growth. Apriori is a basic and common algorithm that begins with single items and accumulates to form larger itemsets. It employs a bottom-up strategy and deletes any itemsets that have a support below the minimum support threshold. While Apriori is straightforward to use, it can be slow and costly in terms of efficiency, particularly if the dataset is very extensive, as it produces a great many potential item combinations and repeatedly scans the full set.

FP-Growth is a newer and quicker algorithm. It produces a compressed data structure called an **FP-tree** (Frequent Pattern Tree) and employs the tree in

**Table 1.7** A Comparison Between Isolation Forest and Mahalanobis Distance

| Aspect | Isolation Forest | Mahalanobis Distance |
|---|---|---|
| Principle | Detects anomalies by randomly partitioning the data and isolating points that require fewer splits | Measures the distance of a point from the data mean using the covariance structure |
| Assumption | Outliers are easier to isolate than normal points | Data follows a roughly multivariate normal distribution |
| Type | Tree-based ensemble method | Distance-based statistical method |
| Input requirements | Works directly on raw features; no strict distribution assumptions | Requires estimation of the mean vector and the covariance matrix |
| Handling of high dimensions | Naturally handles high-dimensional data with random feature selection | Sensitive to covariance estimation; performance degrades in very high dimensions without regularization |
| Interpretability | Provides anomaly scores; tree path lengths can indicate isolation level | Distance value directly interprets how far a point is from the multivariate center |
| Scalability | Efficient for large datasets due to subsampling and tree structures | Computationally heavy for huge datasets because of covariance matrix inversion |
| Best use cases | General anomaly detection in large, complex, or high-dimensional datasets | Outlier detection when the data distribution is approximately Gaussian and the covariance can be reliably estimated |

exploring the frequent itemsets without the need for constructing the entire possible combinations. It is hence far quicker and better for bigger datasets. It is also more complex and more memory-hungry, though, based upon the character of the data [120]. Overall, the Apriori algorithm is a suitable choice for small and medium-sized datasets and for teaching purposes due to its ease of use. The FP-Growth algorithm is favored in real-world applications involving a large volume of data due to its superior performance. The two algorithms are effective in identifying associations that are useful in making sound business decisions. Beyond retail, association rule learning is also employed in web usage analysis, in bioinformatics applications like the discovery of gene interactions, and in fraud analysis. It facilitates the identification of hidden relationships that are not easily discernible and offers useful insights in various areas.

## 1.5 Advanced Unsupervised Learning

**Advanced unsupervised learning** is a crucial field within machine learning algorithms, with a primary focus on discovering hidden elements, reducing high-dimensional data, and facilitating the intelligent organization of unlabeled data. This chapter examines various unsupervised learning techniques, including **Self-Organizing Maps (SOMs)**, which can transform high-dimensional data into low-dimensional, topologically visualized representations. SOMs provide the foundation for introducing complex datasets, which are used especially in fields such as industrial systems, medical imaging, and marketing.

Additionally, it is worth mentioning the approaches in unsupervised representation learning, which include autoencoders, variational autoencoders, and deep clustering frameworks. These methods are particularly important because they are capable of learning abstract features and providing cluster-friendly representations that usually result in better performance than traditional algorithms when handling high-dimensional, sparse, or noisy data. Manifold techniques, such as t-SNE, UMAP, and Laplacian Eigenmaps, can reveal the intrinsic geometries in complex datasets [121].

Clustering has been a challenging problem in unsupervised machine learning, employed to identify latent patterns and subgroupings within datasets. Among the strong and interpretable methods of clustering is the Self-Organizing Map (SOM), or rather a **Kohonen Map,** named after its creator, Teuvo Kohonen. The SOMs are special neural networks capable of transforming a high-dimensional dataset into a low-dimensional (typically two-dimensional) map, while preserving the topological structure of the input space. Comprises a neuron grid, and each neuron is connected to a weight vector, which during training is updated in order to be able to provide a presentation of patterns in input data [122].

Once a data point is input to the SOM, the neuron whose weight vector is most similar to the input may be said to be the Best Matching Unit (BMU). The BMU and its nearby neurons are then updated to better match the input vector. The network self-organizes itself by means of similar inputs that are active for nearby neurons, building coherent groups or clusters.

**Self-Organizing Maps** were originally presented by Finnish professor Teuvo Kohonen in 1982. The old-style neural networks required input labeling; in contrast, SOMs operate under different rules by employing unsupervised learning, adapting them to the input data distribution without external supervision. The approach defines a particular class of artificial neural networks and is applied as a tool for clustering and visualization in exploratory data analysis. The main domains where SOMs are well applied are: industrial instrumentation, pattern recognition, data mining and processing, process control, robotics, telecommunications, medicine applications, optimization, and product management [123]. The spatial locations (e.g, coordinates) of the nodes in the output space are inherent to statistical features contained in the input space. In this respect, SOMs share some similarities with well-established dimensionality reduction techniques such as Principal Component Analysis (PCA) and Multidimensional Scaling (MDS). All three methods reduce the high-dimensional data into a more manageable, low-dimensional form while maintaining the essential structure or relationships among data points. When SOMs are compared to these classical multivariate techniques, SOMs offer several distinct advantages, mainly due to their nonparametric nature.

Construction of SOMs is explained in three procedures:

- Competition—is a process in which output nodes (neurons) in a self-organizing map are competing against each other in order to give a better case of a specific input sample. We could measure a successful representation by a discriminant function, in which the input vector is matched against the weights of every neuron. If we have a connection between them, then we have a match, and it could serve as a successful representation.
- Cooperation—Describes how SOM's topographic structure gathers together information of comparable properties, attempting to reflect natural Human neurons' behavior, where such neurons processing comparable information are found close to one another due to interaction over short synaptic connections.
- Adaptation—The connected neurons and local units in the map's weight vectors undergo adjustment to prefer the larger values of their discriminant functions so that the pertinent nodes become increasingly closer to the input sample.

SOM consists of a grid of neurons, where each neuron is associated with a weight vector of the same dimensionality as the input data. During training, the neuron whose weight vector is most similar to the input (the Best Matching Unit, or BMU) is selected, and its weights, as well as their neighboring weights, are updated to move closer to the input. Over a certain time of training, this process results in topological ordering, resulting in clustering of similar inputs that activate adjacent neurons in the framework. Data that is represented by the m-dimensional input vector, where elements indicate the attribution $x = (x_1, x_2, \ldots, x_n)'$ to the values of a particular dataset. Suppose there is a large difference among the values that were attributed in the data set. In that case, the requirement is to normalize the data in order to avoid the dominance of a particular attribute or subset of attributes, resulting in equal chances for all attributes, and to improve the numerical accuracy [124]. The z-score transformation is the most commonly used technique to normalize and standardize data before the processing step, by converting each element to a standardized score by subtracting the mean and dividing it by the standard deviation, such that in the new distribution, its mean is zero. Its standard deviation is unity, such that this expression is:

$$\left(x_j - \mu_{xj}\right) / \sigma_{xj} \tag{1.24}$$

Max normalization is yet another widely used method in which every characteristic value is divided by the maximum value in a given property, thus scaling values in the range [0 1]. Normalization of unit length is yet another technique, scaling all elements so that the result vector has a unit norm, again, min-max transformation, linearly rescaling data, in a defined range, usually [0 1]. Typical uses of logarithmic transformation in exponential distributions of variables help reduce skewness and compress large values, thereby allowing for research on these variables.

The output layer in a SOM is a low-dimensional representation of input data, a two-dimensional map that consists of columns and rows. The output layer consists of nodes (or neurons), which are in a topologically organized structure, and the maximum clusters are directly controlled by the granularity of clusterization as well as the effectiveness of analysis of a model, which is defined by the maximum number of nodes in the output layer (Fig. 1.31).

In traditional SOMs, a rectangle or a hexagonal structure is typically used to arrange network topology. These configurations differ in whether or not neurons are connected to neighbors, and how, during training, information propagates. For a

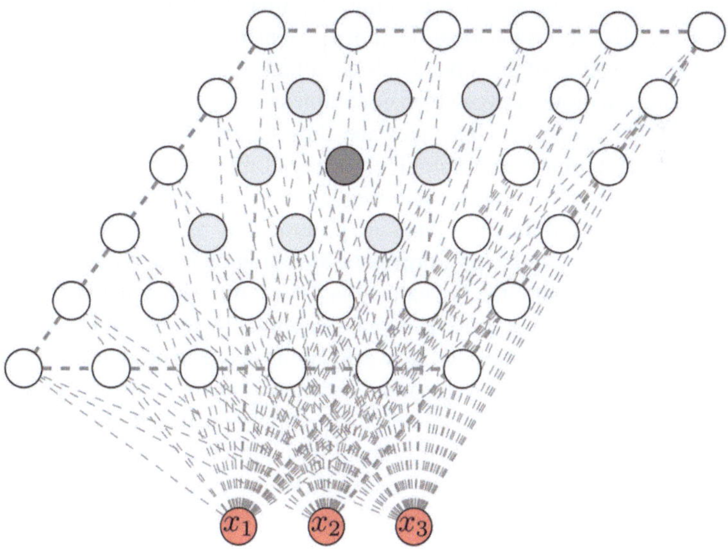


**Fig. 1.31** Self-Organizing Maps (https://commons.wikimedia.org/wiki/File:Self-organizing-map.svg)

rectangular topology, structure is defined in terms of each internal node's link to four neighbors from each, i.e., the above, below, left, and right. For a hexagonal topology, each internal node is connected to six neighbors, in a broadly circular structure based upon each node [113–115]. The hexagonal structure is commonly adopted as a consequence of higher flexibility and larger variability in regional size, which may promote the topological ordering and ability to cluster a SOM.

Initializing a SOM begins with the initialization of the weight vectors, considered as $w_i = (w_{i1}, w_{i2}, \ldots, w_{in})$, where $i = 1, 2, \ldots, n_i$ and n is the number of nodes in the output layer. The weight vectors are the entry points from the input layer to the output layer and are updated in each iteration during the learning phase. Initial dispersion of weights could have a significant effect on the final outcome of the SOM, and so deserves careful consideration.

The initial weights, $w_i(0)$ for $t = 0$, may be chosen either in a random or systematic fashion. One of the most popular ways to initialize weights as small random numbers, such that no initial ordering is induced upon the network at start, is a simple technique. This technique has, however, one essential constraint: the weights so assigned must be sufficiently different from each other in order not to result in consistency in nodes. When initial weights are set close to zero, care must be exercised in terms of adopting a large learning rate and a large neighborhood parameter, at least during the initial phase of learning (usually, 1000 iterations and beyond), to result in effective convergence [125]. Regardless of its popularity, however, random initialization may occasionally result in a slower convergence time.

In order to attenuate this effect, a different method is used to select sample vectors from the training sets in a random fashion in order to act as initial vectors for weights. This method introduces a lot of structure to the space of weights from the onset, and it can really enhance training. Another, higher-level, and systematic method of initialization is the identification of latent vectors that are associated with the largest latent values of the input data's covariance matrix. The vectors of weights are then initialized along a two-dimensional linear subspace spanned by these main components. This approach leverages the data's intrinsic structure, resulting in faster merging and improved topological sorting (Table 1.8).

SOMs are applied in cell manufacturing systems, in particular, in cell designing problems. Groupings of machine parts are promoted by associating groups of parts and machines based on matching prevalent operating characteristics. This leads to efficient cell design and the visualization of relationships between parts and machines, facilitating the optimization of production layout. In marketing and product positioning, SOMs are used to find associations between products and consumer preferences. They enable positioning and segmentation simultaneously by converting brands and consumers' perceptions into a common perceptual space through perceptual mapping analysis [126]. Compared to traditional multivariate approaches, SOMs are not constrained by strong distributional assumptions and are efficient in handling heterogeneously scaled data and missing values.

SOMs, in fact, have been demonstrated to have value in chemistry as well as in medical imaging applications, where we deal with complex, high-dimensional data structures. For chemistry, the combinatorial design of cannabinoid compounds is an

**Table 1.8** The equations applied in Self-Organizing Map algorithms

| Step | Purpose | Equation | Explanation |
|---|---|---|---|
| Input representation | Represent each input vector | $x = [x_1, x_2, \ldots, x_n]$ | Input is an n-dimensional feature vector presented to the map |
| Distance calculation | Compute the similarity between the input and each neuron | $d_j = \|\|x - w_j\|\|$ | Euclidean distance between input xx and the weight vector $w_j$ of neuron j |
| Competitive learning | Select the best matching unit (BMU) | $c = \text{argmin}_j d_j$ | The neuron cc with the smallest distance is the winner |
| Weight update | Adapt neuron weights toward input | $w_j(t + 1) = w_j(t) + \eta(t)\, h_{cj}(t)\, [x - w_j(t)]$ | Learning rate $\eta(t)$ and neighborhood function control update magnitude |

example where SOMs have been applied. The SOMs in such cases are mapped to a hexagonal grid such that the output layer of the SOM is shown, so that researchers may effectively see and explore the chemical space.

Medical imaging and diagnostics utilize SOMs to detect disorders from various modalities of medical images, including electrocardiograms (ECG), computed axial tomography (CAT) scans, and ultrasound scans [127]. Image segmentation is a common application where SOMs are used to aid in identifying regions of interest, which are essential for diagnosis. They can distinguish between anatomical traits and abnormal ones, thereby increasing diagnostic efficacy and helping doctors in treatment procedures.

While supervised learning has shown impressive progress in deep learning, most prominently in classification, its foundation is nonetheless based on unsupervised learning due to its limitation in applying only to labeled data. Unsupervised learning enables the discovery of meaningful representations, which are applicable in downstream applications such as clustering, feature reduction, and classification. Various structures have come to light in an attempt to enable unsupervised learning of representations, such as:

- **Autoencoders**, an artificial neural network that is used to learn the efficient codings of unlabeled data, by transforming the input data and recreation of input data from the encoded representation
- **Deep Boltzmann Machines** are a generative graphical model, i.e., a stack of a few layers of hidden units, where there are edges between units from different layers, but from the same layer, there are no such links. These are utilized to learn hierarchical feature representations from raw data, where the output of a layer serves as input for the next layer. The algorithm is layer-wise pre-trained using a Markov random field and then fine-tuned by the backpropagation technique

- **Deep Belief Networks** it is comprised of several layers of neurons connected to later layers, and no intra-layer links. Whereas, in training, each layer is a different model, dependent upon the output of the previous.
- Variational Autoencoders are a type of artificial neural network that produces data close to the input used in training, whose essential objective is denoising.
- **Generative Adversarial Networks** it is applied in training two neural networks to play against one another in a bid to develop increasingly realistic new data from a provided set of training data.

The methods are categorized into three groups:

1. Autoencoders—Models focus on learning identity-preserving transformations.
2. Sparse models—Like sparse autoencoders, so that there are restrictions on the learned code.
3. Generative models—VAEs and GANs, that directly model the generation procedures of data.

The majority of deep models rely heavily upon sets of large, labeled datasets, wherein a very small fraction of unsupervised techniques are applied [128]. This approach could then be demonstrated in two groups of limitations, which restrict their applications in higher or less visually interpretable domains:

- The necessity for vast quantities of human-annotated data makes such models costly and laborious to create.
- Regardless of how well such models work for particular tasks, however, such models are frequently unable to deliver thoughtful comprehension of the problem domain at its core; the deep insight that such models provide is typically only for viewing dominant input features.

In recent years, there has been a significant interest in node-embedding techniques, resulting in a wide variety of methods that result in methodological diversity. To proceed with this method, the unified encoder-decoder framework is utilized, providing a consistent structure for understanding. This type of framework revolves around two components:

- **Encoder** is the mapping function, from transforming each graph node to ENC: V –> $R^d$ low-dimensional vector representation (embeddings) $z_i \in R^d$, where zi is the embedding of graph node $v_i \in V$.
- **Decoder** is the function that applies pairs of node DEC: Rd. x Rd. –> $R^+$ embeddings to transform the real-valued similarity measure of nodes, providing us with structure (Fig. 1.32).

Almost all of these node embedding methods apply so-called shallow embedding methods, in which each node from a graph is directly mapped to an idiosyncratic vector, without a deeper network or a model playing a role in producing such embeddings [129]. Shallow methods of embedding may be broadly classified along two lines [130, 131]:

- Matrix factorization—trying to learn node embeddings from the similarity measure.

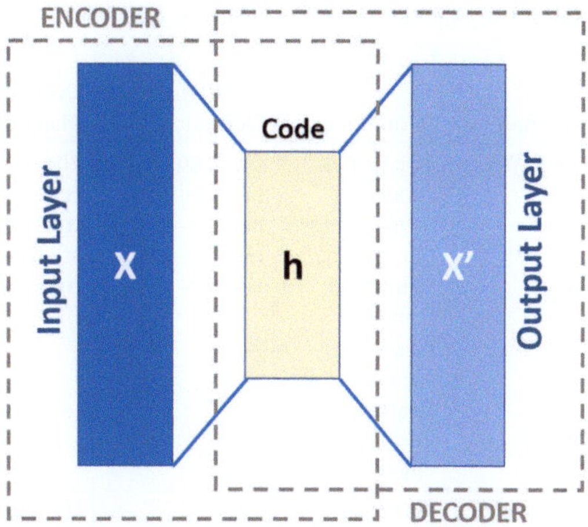


**Fig. 1.32** Autoencoder structure (https://commons.wikimedia.org/wiki/File:Autoencoder_schema.png)

- Laplacian Eigenmaps seeks to link nodes nearby in the embedding space
- Factorization of a graph is a factorization of its adjacency matrix.
- GraRep introduces higher-order relations in terms of providing multi-step links between nodes
- HOPE is more flexible, and it allows a wide range of similarity measures
- The random walk approaches, i.e., DeepWalk and node2vec, have recently been applied, and rather than static indicators of similarity, random walks over the graphs are utilized to learn nodes that are most likely to co-occur close to one another, by generating short random walks, and see which nodes cluster together most of the time (Table 1.9).

**Autoencoders** are designed to learn efficient codings of data by training the network to modify its input. The process consists of two major components: encoders whose function z = f(x), used to map the input x into a latent representation z; and a decoder whose function x = g(z). In both cases, the f and g are usually used by a neural network composed of one or more non-linear layers [112]. The mean squared error (MSE) represents the objective function that minimizes reconstruction loss during training.

$$\mathcal{L}(x,\hat{x}) = \|\hat{x} - x\|^2 = \|g(f(x))\|^2 \tag{1.25}$$

This loss prompts the autoencoder to preserve useful information in the input and to get rid of information that is not useful.

To prevent the autoencoder from learning the identity function, the ability to copy input as output without abstraction. The most common restriction is to limit the size of the latent representation z, and these kinds of models are referred to as **undercomplete autoencoders**. In the process of compressing the input into a smaller latent space, the autoencoders are forced to learn the important features that can reconstruct the input with minimal error [132]. Suppose an autoencoder is linear and trained under MSE loss. In that case, it can learn the same subspace as PCA, where the subspace is spanned by the hidden units in the same linear subspace as the

**Table 1.9** Shallow embedding methods overview

| Method | Core Idea | Similarity Definition | Decoder Mapping | Symmetry of Similarity and Decoder | Key Notes |
|---|---|---|---|---|---|
| DeepWalk | Generates node embeddings by treating truncated random walks as sentences and training a skip-gram model | Co-occurrence probability of nodes within a random walk window | The dot product between node embeddings predicts walk co-occurrence | **Not symmetric**: The probability of node $u$ given $v$ in a walk is not identical to $v$ given $u$ | Learns representations that preserve network proximity |
| node2vec | Extends DeepWalk with biased random walks that interpolate between breadth-first and depth-first search | Same as DeepWalk, but with adjustable walk bias parameters p and q | Dot product predicts co-occurrence in biased walks | **Not symmetric** for the same reason as DeepWalk | Balances structural and community information |
| LINE | Preserves first-order (edge) and second-order (neighbor distribution) proximity using explicit objective functions | First order: Edge probabilities; second order: Conditional neighbor distributions | Separate embeddings for source and target roles in edges | Source and target embeddings make the similarity–decoder relationship **asymmetric** | Supports large-scale graphs efficiently |
| Matrix factorization (e.g., Laplacian Eigenmaps, GraRep) | Factorizes adjacency or similarity matrices | Directly derived from the graph Laplacian or higher-order matrices | Reconstruction of matrix entries via symmetric dot products | **Symmetric**: Similarity definition matches decoder inner product | Provides clear theoretical connections to spectral clustering |

input data's principal components, showing the capability of autoencoders to generalize classical linear approaches [133, 134].

**Sparse Autoencoders** employ overcomplete representations, which place constraints that favor sparsity in the activation, where the latent space is high. However, only a small subgroup of neurons is active for any input, which causes these representations to become stronger and interpretable [135]. Another popular technique for inducing sparsity is to add a regularization term in the loss function that punishes departures of activation of a hidden unit from a small desired mean activation. For a dataset of n samples, $\hat{\rho}_j$ of a hidden unit j:

$$\hat{\rho}_j = \frac{1}{n}\sum_{i=1}^{n} z_j\left(x_i\right) \tag{1.26}$$

where activation of latent unit j for input x is defined as $z_{j,}$ where $\hat{\rho}_j$ is contrasted against the target sparsity level of ρ, and to ensure that, a Kullback-Leibler (KL) divergence penalty is added $\hat{\rho}_j \sim \rho$ to loss in reconstruction as originally defined:

$$\mathcal{L}\left(x,\hat{x}\right) = \left\|\hat{x} - x\right\|^2 + \lambda\sum_{j=1}^{m} KL\left(\hat{\rho}_j \middle\| \rho_j\right) \tag{1.27}$$

λ is a regularization hyperparameter, and m is the number of hidden units.

**Denoising Autoencoders** (DAEs) introduce a robust strategy for representation learning by incorporating stochastic corruption during training. Unlike standard autoencoders that learn to reconstruct the input x, the DAEs are trained to reconstruct the original inputs from a corrupt version $\overline{x}$, which is forcing the network to learn invariance, high-level features that generalize beyond simple memorization of input data:

$$\mathcal{L}\left(x,\hat{x}\right) = \left\|g\left(f\left(\hat{x}\right)\right) - x\right\|^2 \tag{1.28}$$

where $\hat{x}$ is a stochastically corrupted copy of x, and f( $\hat{x}$ ) represents noisy input and g() decodes it for input reconstruction.

DAEs are capable of learning robust representations that improve performance in downstream tasks, i.e., classification and clustering, in terms of being able to undo the corruptions. DAEs are generative models as well in that, after corruption and reconstruction iteratively, they come to learn something about the elements of the distribution of the original data [137]. Another variation of DAEs is Dropout Autoencoders, a combination of input corruption and dropout regularizer applied in the hidden layers, where in each pass during learning, a random subset of neurons is "dropped out" such that there is repetition as well as robustness in learned features.

**Contractive Autoencoders** (CAEs) are correcting the model's sensitivity to small disruptions of inputs, which imposes direct regularization on the encoder's gradient, presented by:

$$\mathcal{L}\left(x,\hat{x}\right) = \left\|g\left(f\left(\hat{x}\right)\right) - x\right\|^2 + \lambda\sum_{i=1}^{m}\left\|\nabla_x z_i\right\|^2 \tag{1.29}$$

$z_i$ represents the activation of the i-th hidden unit, the $\nabla_x z_i$ is the Jacobian of the encoder output with respect to input x, and λ controls the strength of contraction. The Frobenius norm of the λ Jacobian is the added penalty term, which encourages the network to map the inputs that are near to similar latent representations, leading to smooth latent spaces and robust feature learning, which can be valuable in noisy or high-dimensional data environments [138, 139].

**Variational Autoencoders** (VAEs) are probabilistic models that not only compress data but also learn the underlying probability distribution of the data, enabling tasks such as data generation, anomaly detection, and semantic interpolation in the latent space [140, 141]. In place of encoding an input x in a unique determining point z, predicting distribution over latent variable, usually through a multivariate Gaussian, then the procedure is where the decoder samples a latent vector z from the distribution to reconstruct x. Training of VAEs minimizes the **Evidence Lower Bound** (ELBO) of the marginal log-likelihood of data:

$$\log p(x) \geq E_{q\phi(z|x)}\left[\log p_{\theta}(x|z)\right] - D_{KL}\left(q\phi(z|x) \| p(z)\right) \tag{1.30}$$

$E_{q_s}$ presents the reconstruction term, $D_{KL}(q_{\Phi}(z|x)||p(z))$ is the regularization term. This process has two distinct goals:

- Precise reproduction of input
- Regularized latent space encouraging smoothness and meaningful interpolation.

**Deep clustering** is a newly emerging field of machine learning methods that leverages the representational ability of deep learning and classical clustering methods to enable better research into challenging clustering problems for high-dimensional, complex data. Whereas traditional clustering must adhere to a few universally defined rules, deep clustering learns a different feature space in which intra-cluster similarity and inter-cluster diversity are optimized in particular. With the use of neural networks, such as autoencoders or CNNs, it is possible to map data into a latent space where clusters are more distinguishable and manageable [142]. The deep clustering model learns to tailor the representation to facilitate clustering better, even if the initial data we have isn't clusterable.

Deep clustering introduces a novel similarity metric by transforming data at training time, such that similar items (same image category, similar gene expression patterns) are clustered together. Not similar items are distanced, as opposed to traditional clustering, where we would have used static distance measures such as **Euclidean, Manhattan or cosine distance**, which are pre-specified. This novel model also facilitates the easier interpretation of data, even if the original raw data itself might not be ideal, being too raw and complex. In contrast, deep clustering can complement visualization methods such as t-SNE or UMAP, revealing more structured patterns. Contrary to traditional clustering approaches that are applied directly to raw data, deep clustering progresses through hierarchical, non-linear feature mappings that capture the data structure more effectively, thereby facilitating better clustering. This combination has proven highly effective in applications such as image analysis, text mining, speech recognition, and bioinformatics, where raw input data is typically noisy, high-dimensional, and lacks explicit labels.

Deep clustering approaches may be categorized into four frameworks: single-view, multi-view, semi-supervised, and based on transfer learning. The most researched are **single-view deep clustering** approaches, which entail learning latent representations from a single source of data using deep architectures such as autoencoders or convolutional neural networks, and then clustering in this learned space. **Multi-view deep clustering** works with different forms of input data (or modalities), e.g., when one fuses genomic data with clinical data to learn a richer feature representation that captures variable aspects of the same entities. **Semi-supervised deep clustering** involves leveraging minimal label information to direct a clustering process, allowing a model to take advantage of both labeled and unlabeled data, a significant factor in applications such as bioinformatics, where labeled data may be costly or inaccessible. Transfer learning-based clustering involves transferring learned features from one domain to another, thereby facilitating clustering in target domains where ample training data is scarce.

**Single-view deep clustering** is the most widely used form of deep clustering methods. The methods are rooted in a single modality or source of data and seek to learn feature representations while simultaneously clustering. Autoencoders, deep autoencoders, or convolutional neural networks are learned to compactify the high-dimensional input into a lower-dimensional latent space. Cluster objectives, often based on loss operations such as k-means distance or soft assignment likelihoods, are optimized alongside reconstruction loss during learning. This joint learning enables the network to learn feature embeddings crafted specifically to encourage cluster separation. Single-view methods such as **Deep Embedded Clustering (DEC) and Deep Clustering Network (DCN)** illustrate this approach. These methods are most useful in situations where the dataset is homogeneously derived from a single data type, such as gene expression levels, image pixels, or EEG signals [143, 144]. The utility of single-view deep clustering is in its simplicity as well as its strength in situations where raw data is high-dimensional and not easily discriminable using traditional clustering methods.

**Multi-view deep clustering** generalizes the single-view technique by incorporating multiple heterogeneous data sources or modalities that describe the same entities. For instance, in biomedical applications, a patient might have gene expression profiles, radiological image data, and clinical text records. Each of these views represents different facets of the underlying biological system [145]. Multi-view deep clustering models process such divergent inputs either separately in parallel neural branches or together in a fused architecture in order to learn a shared low-dimensional representation. This aggregated feature space is then employed in clustering. The main objective is to take advantage of the complementarity of information from various data views in order to boost clustering robustness and accuracy. Difficulties with this technique include aligning and weighing different modalities effectively, handling missing views, and view dominance, where a particular modality overweights the clustering result. Nevertheless, multi-view clustering has demonstrated promising potential in applications such as cancer subtype classification, where a fusion of genomic and clinical data enhances the biological relevance of the clusters. Preserving view-specific as well as shared information, multi-view deep clustering represents a richer and richer structure compared to single-view methods (Table 1.10).

**Table 1.10** Summary of Single-View Deep Clustering

| Method | Key Idea | Network Architecture | Clustering Objective | Representative Loss Functions | Strengths | Limitations |
|---|---|---|---|---|---|---|
| DEC | Learn a latent space and cluster assignments jointly | Autoencoder (encoder used after pretraining) | Minimize the KL divergence between soft cluster assignments and a target distribution | KL divergence loss + reconstruction loss during pretraining | Simple pipeline, widely used baseline | Sensitive to cluster number and initialization |
| VaDE | Combine a variational autoencoder with a Gaussian mixture model in latent space | Variational autoencoder | Maximize evidence lower bound with a Gaussian mixture prior | VAE reconstruction + KL divergence to mixture prior | Handles complex latent distributions, probabilistic clustering | Higher computational cost, requires careful tuning |
| DCN | Integrate k-means clustering with autoencoder reconstruction | Autoencoder with joint optimization | Minimize the k-means objective in the latent space while preserving reconstruction | k-means loss + reconstruction loss | Encourages discriminative features while maintaining input fidelity | Balancing clustering and reconstruction losses can be tricky |
| IMSAT | Encourage invariance to perturbations and maximize mutual information | Feed-forward neural network | Maximize mutual information between inputs and predicted clusters | Mutual information loss + consistency loss | No need for pretraining, robust to input noise | May collapse to trivial solutions without careful regularization |
| DAC | Treat clustering as a binary pairwise classification of similarity | Convolutional neural network | Predict whether pairs of samples belong to the same cluster | Binary cross-entropy on pseudo labels | Effective for image clustering, it learns similarity directly | Sensitive to pseudo-label quality |

**Semi-supervised deep clustering** lies between supervised and unsupervised learning by adding a minimal level of labeled data to the clustering. For a majority of real-world problems, such as genetics, medicine, and neuroscience, gathering labeled data is expensive and time-consuming, whereas enormous quantities of unlabeled data are available. Semi-supervised deep clustering bridges this gap by exploiting well-known labels to give direction to the learning of the feature space and molding of clustering boundaries [16]. This is typically done by adding extra loss terms (e.g., supervised classification loss) in addition to the unsupervised clustering objective during training. The labeled samples direct the network in balancing its latent space along meaningful class partitions, in effect refining clustering quality for the unlabeled samples. This is extremely effective when there is high-quality labeled data accessible only for a subset of conditions or classes, such that constituent models learn how to generalize and interpret novel patterns in a majority of unlabeled samples [146]. For instance, in biomedical data analysis, a few labeled disease types may facilitate the clustering of novel subtypes, providing significant information for diagnosis or drug discovery. Semi-supervised clustering is, therefore, an effective tool in situations where partial supervision is accessible.

**Transfer learning-based deep clustering** uses pre-trained models or previously learned feature representations in a prior domain in a new, often less well-resourced target domain. This is particularly useful in applications where computational resources or labeled data are limited in the target context. The idea is to utilize knowledge learned from a large, possibly labeled source dataset to initialize a neural network or to prescribe a latent space for clustering in a target dataset. For instance, a pre-trained model from a well-annotated medical imaging data set may be reused for clustering another, small-scale data set with similar imaging profiles [147, 148]. This method often involves fine-tuning a pre-trained model using loss functions for clustering or adapting the latent space through domain adaptation procedures. Transfer learning is particularly useful in bioinformatics and biotechnology, where pre-trained gene expression models or protein embedding models can be reused for correlated clustering tasks, such as identifying disease subtypes or describing drug response profiles. Through the reduction of needs for full-scale retraining and labeled sets, transfer learning-based deep clustering reduces deployment time and improves performance in practical applications, specifically in cross-domain biomedical research (Table 1.11).

When discussing specific techniques used in deep clustering, multiple approaches are available. Beginning with DEC, which integrates representation learning and clustering into a unified framework. DEC begins by using a stacked autoencoder to initialize the network and generate compact, low-dimensional embeddings from high-dimensional data. Instead of treating clustering as a separate step after feature extraction, DEC incorporates a clustering objective into the training process by defining a soft assignment of data points to clusters using a Student's t-distribution. The model iteratively refines both the feature space and cluster centroids by minimizing the KL divergence between the soft assignments and an auxiliary target distribution that emphasizes confident cluster assignments. This iterative

**Table 1.11** Methods used in Multi-View Deep Clustering

| Method | Key Idea | Network Architecture | Integration Strategy | Loss Functions | Strengths | Limitations |
|---|---|---|---|---|---|---|
| DCCA | Learn nonlinear transformations of each view so their latent representations are maximally correlated | Separate deep networks for each view | Correlate view-specific latent spaces | Canonical correlation objective | Captures nonlinear shared structure across views | Requires paired data and may ignore view-specific information |
| MV-DEC | Extend DEC to jointly learn embeddings and cluster assignments across multiple views | Autoencoders per view with shared cluster layer | Concatenate or fuse embeddings from all views | KL divergence loss + reconstruction loss | Exploits complementary information from each view | Sensitive to the quality of view fusion and cluster number |
| DMVSC | Integrate spectral clustering with deep feature learning | Deep encoders for each view | Shared graph Laplacian and co-regularization between views | Graph Laplacian loss + reconstruction loss | Preserves graph structure and cross-view consistency | High computational cost for large graphs |
| Co-AE | Train autoencoders for each view with a co-training regularizer that enforces agreement | One autoencoder per view | Minimize reconstruction error while maximizing agreement | Reconstruction loss + co-training consistency loss | Balances view-specific and shared information | May underperform if one view is noisy or missing |
| JMVC | Learn a common latent space where clusters from different views align | Shared encoder with view-specific decoders | Align latent representations and cluster assignments | Clustering loss + view alignment loss | Handles heterogeneous views and missing data | Complex training and parameter tuning |

optimization yields embeddings that are highly discriminative for clustering, enabling the model to adaptively reshape the latent space and enhance cluster separability.

Next, **Variational Deep Embedding (VaDE),** a deep unsupervised clustering algorithm that combines representation learning and clustering in a framework of generative probabilistic models. VaDE is an extension of the **traditional Variational Autoencoder (VAE)**, where a **Gaussian Mixture Model (GMM)** is used as a prior in latent space. This enables a model to learn a structure of the data as independent clusters, in addition to continuous latent representations. For a conventional VAE, latent variables are defined as a single Gaussian, which limits its ability to capture multi-modal or cluster-like distributions. VaDE substitutes this assumption for a compound of Gaussians, allowing it to associate each data point, in an explicit manner, with a cluster from latent space. In training, the model is optimized against a variational lower bound over data likelihoods using stochastic gradient descent and reparameterization tricks, and learns both the parameters of the GMMs and the neural networks.

Following this, we have **DeepCluster**, a deep unsupervised learning method that aims to discover meaningful visual features from unseen data by integrating CNNs and iterative clustering. DeepCluster suggests a fresh training procedure where the network learns to represent things by switching back-and-forth between two core steps: (1) feature extractions from data by a CNN, and (2) feature clustering by a classical algorithm such as k-means. The cluster indicators are then perceived as pseudo-labels, serving to refine the CNN's parameters in a supervised manner. This closed-loop operation is applied and gradually improves the feature space such that semantically meaningful and coherent clusters are obtained. Contrary to pretext-task-based approaches (such as rotation prediction or jigsaw puzzles), DeepCluster uses structure uncovered by clustering to directly guide visual representation learning. Even though no ground-truth labels are used, the approach yields competitive performance in downstream tasks, such as image classification and object detection, particularly when fine-tuned from small labeled sets. DeepCluster is particularly effective in applications such as computer vision and radiology, where there are few annotated samples available, yet there are large amounts of available samples. By effectively converting unsupervised clustering into a self-supervised feature learning tool [149], DeepCluster closes the gap between data grouping and representation learning, yielding a scalable and label-efficient substitute for classical supervised training (Table 1.12).

**Joint Unsupervised Learning (JULE)** is a deep clustering approach that integrates representation learning and clustering in a closed-loop, recurrent process. JULE centers on iteratively refining both the feature representations as well as the clustering structure by alternating between the two tasks. The insight is to start off from agglomerative clustering, a type of hierarchical clustering where each data point is initially in its very own cluster, and then progressively merges pairs that are nearest to each other. Following every merging step, the temporary cluster assignments are employed as pseudo-tags to train a convolutional neural net, refining the quality of learned features. The better features are then used to guide the next

**Table 1.12** Summary of Semi-Supervised Deep Clustering methods

| Method | Key Idea | Network Architecture | Use of Labeled Data | Loss Functions | Strengths | Limitations |
|---|---|---|---|---|---|---|
| Deep semi-supervised clustering (DSSC) | Combine reconstruction of inputs with a clustering objective and small label guidance | Autoencoder with shared encoder | A small labeled subset guides cluster formation | Reconstruction loss + clustering loss + cross-entropy on labeled samples | Improves cluster quality even with very few labels | Sensitive to label noise and balance between losses |
| Semi-supervised deep embedded clustering (SS-DEC) | Extend DEC by adding a supervised classification branch | Autoencoder encoder + classification head | Labeled samples provide explicit class signals during training | KL divergence clustering loss + classification loss | Stable optimization and better cluster-class alignment | Requires careful weight tuning between supervised and unsupervised terms |
| Deep constrained clustering network (DCCN) | Incorporate pairwise must-link and cannot-link constraints | Convolutional encoder network | Labeled pairs specify whether two samples should share a cluster | Pairwise constraint loss + reconstruction loss | Flexibly encodes domain knowledge through constraints | Relies on accurate constraints; cost increases with many pairs |
| Hybrid self-training clustering (HSTC) | Iteratively refine pseudo-labels with a small set of ground-truth labels | Encoder network with iterative label update | Seed labels start pseudo-label generation | Cross-entropy with pseudo-labels + consistency regularization | Gradually improves clusters and classifier jointly | Pseudo-label errors can propagate if early predictions are poor |
| Semi-supervised Variational deep embedding (SS-VaDE) | Combine a variational autoencoder with a Gaussian mixture prior and labeled supervision | Variational autoencoder | Labeled subset guides mixture component assignments | VAE reconstruction + KL divergence + supervised classification | Probabilistic framework handles complex distributions | Computationally heavy and sensitive to initialization |

clustering step, resulting in improved groupings. This back-and-forth approach continues until a whole hierarchy of data points is established, effectively instructing the CNN to learn a semantically rich representation without requiring a single labeled image. JULE differs from approaches such as DEC and DeepCluster in employing a hierarchical, rather than flat, type of clustering and in tightly integrating the learning of representations with the structure of clustering itself. JULE is most effective in image clustering problems where visual similarity is to be learned from scratch, such as face grouping or object classification. JULE's main strength is to prove that clustering and learning of representations could evolve together, yielding much better performance than training them separately. This makes it a foundational approach for deep clustering pipelines aimed at simulating human-like perception through structure iteration and meaning refinement.

**ClusterGAN** is a deep generative clustering method introducing a novel extension of the Generative Adversarial Network (GAN) architecture to facilitate unsupervised clustering in the latent space. ClusterGAN defies a general assumption in vanilla GANs that the latent space is a continuous Gaussian or uniform distribution by adding discrete latent variables to continuous ones. The latent vector is specifically formed from a combination of a one-hot categorical random variable (to embody cluster identity) and a continuous noise source. This hybrid latent space architecture enables the generator to create samples from different clusters, while a discriminator and auxiliary encoder work together to ensure that both reconstruction and clustering losses are met [125]. The encoder projects the generated samples back to their initial latent codes, and the entire system is trained adversarially to minimize both reconstruction loss and adversarial loss simultaneously. What ClusterGAN is most remarkable for is its capability to create high-quality fake samples affiliated with particular groups — something most conventional clustering methods are not able to accomplish. Different from other deep clustering methods based upon feature extractions and then clustering, ClusterGAN integrates data generation, clustering, and feature discovery into a unified whole architecture. ClusterGAN excels in particular over image databases such as MNIST and CIFAR-10, and is well suited to applications where sample generation as well as cluster discovery are required — such as in synthetic biology, data augmentation, or unsupervised domain modeling [112]. ClusterGAN's hybrid latent space architecture represents a groundbreaking improvement in deep clustering, bridging the distance between unsupervised discovery and generative modeling.

**Manifold learning** is a higher-level unsupervised learning technique that is primarily interested in discovering and determining intrinsic low-dimensional structures (manifolds) that are hidden in high-dimensional data sets. This is as opposed to traditional linear methods, such as PCA, in terms of "perceiving" data points themselves. The data points are assumed to lie on curved, non-Euclidean surfaces in a higher-dimensional space. The intuition behind this is rooted in the fact that even though data seems to increasingly become complex when in a higher-dimensional space, in a majority of cases, it is reduced to a lower-dimensional shape, which is not as hard to comprehend. This approach aims to reduce and diminish dimensionality while retaining significant geometric aspects of the data. Some algorithms are concerned primarily with retaining global distances between data, whereas others are interested in retaining relations between nearby points. Relevant algorithms that we shall reference are Isomap, which keeps global pairwise distances; Locally Linear Embedding (LLE), which keeps local neighborhood structure; Laplacian Eigenmaps; and t-SNE, which looks to preserve either connectivity or probability distributions [131]. More recently, however, UMAP has been used more frequently than t-SNE, as it is faster when dealing with similar data structures. Each of these techniques has distinct advantages and is applied to different data structures.

Whenever we consider applied use cases for manifold learning, it is heavily employed for the visualization of data, the recognition of patterns, and the determination of similarity measures for points in data. This is particularly useful in applications such as face recognition, where recognizing subtle changes in high-dimensional image data is crucial for effective operation. In genetics, bioinformatics, and omics, manifold learning is applied in gene-disease association prediction, feature selection for cancer classification using genetic algorithms, research on protein interactomes and functional linkages across multiple gene networks, and studies on the structure of human populations, among other applications. Manifold learning has undergone significant developments with the introduction of deep generative models, including Variational Autoencoders and Generative Adversarial Networks. Not only are these models able to learn and investigate low-dimensional data, but they are also able to create fresh, realistic samples from such spaces. This facilitates a sort of double duty in that not only are we able to comprehend the structure of data, but we are also able to create fresh content based upon learned patterns. Greater insight into manifold learning is enhanced by a mathematical and geometric perspective. Manifold learning takes as a starting point that most high-dimensional data is contained in or close to a smooth, low-dimensional manifold in the high-dimensional ambient space. This approach is connected to disciplines such as spectral graph theory and differential geometry, where algorithms attempt to approximate the geometry of such manifolds in terms of discrete structures, such as neighborhood graph sequences. Algorithms such as Laplacian Eigenmaps construct similarity graphs of local regions and compute embeddings in terms of graph Laplacians, where the original, underlying geometry of the manifold is retained. Different approaches, such as Isomap, take geodesic distances in an effort to capture underlying manifold structure, whereas LLE derives local linear structure by representing each point as a linear combination of neighbors. These are viewed as a nonlinear extension of traditional techniques, such as multidimensional scaling, to better accommodate data with curved or nonlinear structures.

Recently, fast manifold learning methods, such as UMAP, have gained popularity due to their speed and ability to preserve structure. UMAP constructs a rough topological structure of the high-dimensional space and tries to preserve that structure in the lower-dimensional embedding via optimization. In contrast to approaches such as t-SNE, UMAP performs better in preserving both local and global relations in data, and scales more effectively to larger datasets, making it applicable to practical problems like single-cell genomics, image registration, and large-scale language embeddings.

In this approach, however, a great deal of difficulties remain. Manifold learning methods are usually challenged by sparse, noisy, or high-dimensional streaming data. Overcoming such limitations involves integrating probabilistic graphical models, manifold regularization, and spectral embedding methods. These methods enhance the robustness and scalability of the approach, particularly in applications such as genetics, systems biology, and social networks, where relationships are intricate and often only partially observed. Increasing research is examining how manifold-based representations can be applied in broader contexts, such as semi-supervised learning, transfer learning, and graph neural networks, thereby broadening their applicability in various machine learning tasks.

Evaluating unsupervised learning algorithms is a challenging task, primarily due to the lack of specific labels, which prevents an objective measurement of the model's performance. While in supervised learning, where prediction is directly compared against known outcomes, unsupervised methods, such as clustering algorithms, must be indirectly evaluated. Evaluation measures are usually divided into internal and external measures. Internal measures evaluate cluster quality based on internal data properties without any external information. These are such measures as the Silhouette Score, which determines how well an object belongs to its own cluster in contrast to other clusters, the **Calinski-Harabasz Index**, which quantifies the between-cluster dispersion ratio over a cluster dispersion, and the Davies-Bouldin Index, which determines cluster compactness and separation. These internal measures give highly computationally efficient clustering performance estimates and are applicable when no labeled data is available. Sometimes, however, they may misrepresent cluster quality, particularly when different-shaped, -density, or -scale cluster shapes are encountered.

External evaluation measures are applied when there is some available labeled data (they are usually only available for benchmarking purposes), so there is a direct evaluation of a clustering. Some of these are the **Adjusted Rand Index (ARI),** which adjusts for chance groupings when comparing cluster results against known labels; **Normalized Mutual Information (NMI),** which quantifies the mutual information between cluster assignments and true classes; and Purity, which is a measure of how well each cluster has a single class in it. While external measures are useful in comparing unsupervised cluster results against known classification problems, they are subject to bias from label distribution as well as class imbalance. Relative evaluation highlights the value in comparing several different algorithms or parameter settings under equivalent conditions, as opposed to dependent absolute metric scores [129, 132]. This is significant as different measures typically give inconsistent opinions. Evaluation of unsupervised models is muddied by complications such as irregular shapes in the clusters, variations in density, and scalability limitations, causing standard measures to act in unpredictable ways. A multi-metric approach, where internal and external criteria are complemented by qualitative visualization (such as dimensionality-reduced embedding), is typically deemed better for trustworthy evaluation. This multi-metric approach to validating the algorithm ensures an assessment that includes not only the statistical properties of the output clustering but also how well it aligns with domain-specific variants.

## 1.6 Perceptrons and Multilayer Perceptrons (MLPs)

The origins of neural network research lie in an attempt to mechanize the way biological brains process information. In the mid-twentieth century, scientists began to wonder whether machines could not only compute, as earlier digital computers had shown, but also *learn*. This question was shaped by developments in neuroscience, particularly the idea that the brain operates through vast interconnected networks of simple units called neurons.

The first formal step came with the **McCulloch–Pitts neuron** proposed by Warren McCulloch and Walter Pitts in 1943. They described a simple binary threshold unit: a neuron receives multiple binary inputs, sums them, and fires (outputs 1) if the sum exceeds a threshold. This was a profound abstraction—reducing the biophysiological complexity of neurons to a clean logical model. However, their model lacked a mechanism for learning [133, 134].

That missing piece arrived in 1958 with Frank Rosenblatt's **Perceptron**. Working at Cornell Aeronautical Laboratory, Rosenblatt sought to design a machine that could learn to recognize visual patterns from examples, without explicit programming. His perceptron combined a McCulloch–Pitts-style neuron with an adaptive learning rule for adjusting connection weights. It was implemented both as a theoretical model and as a physical machine (the Mark I Perceptron), which used an array of photocells as inputs and motors to adjust weights.

This achievement generated enormous optimism. The perceptron was widely portrayed as a harbinger of machines that could see, think, and learn like humans. The New York Times declared that the perceptron "will be able to walk, talk, see, write, reproduce itself and be conscious of its existence." This enthusiasm was premature—but it marks the perceptron as the first concrete instantiation of machine learning [135].

At its core, the perceptron is a **binary linear classifier**. It takes a vector of real-valued inputs $x = [x_1, x_2, \ldots, x_n]$, computes a **weighted sum** of these inputs, adds a **bias** term b, and then applies a **step activation function** to produce a binary output $y \in \{0,1\}$

$$y = f(wx + b)\, where\, f(z) = \begin{cases} 1 if\ z \geq 0 \\ 0 if\ z < 0 \end{cases} \tag{1.31}$$

Here, $w = [w_1, w_2, \ldots, w_n]$ represents the learnable weights. The bias term b allows the decision boundary to shift away from the origin. Geometrically, the perceptron learns a **hyperplane** that divides the input space into two half-spaces: one labeled positive, the other negative (Fig. 1.33).

The perceptron learns through **iterative weight updates** guided by classification errors [136]. Given a training set of labeled examples $\{(x^{(i)}, t^{(i)})\}$ where $t^{(i)} \in \{0,1\}$, the perceptron processes each input and computes its output $y^{(i)} = t^{(i)}$, no update is made. If the prediction is wrong, the weights are adjusted as follows:

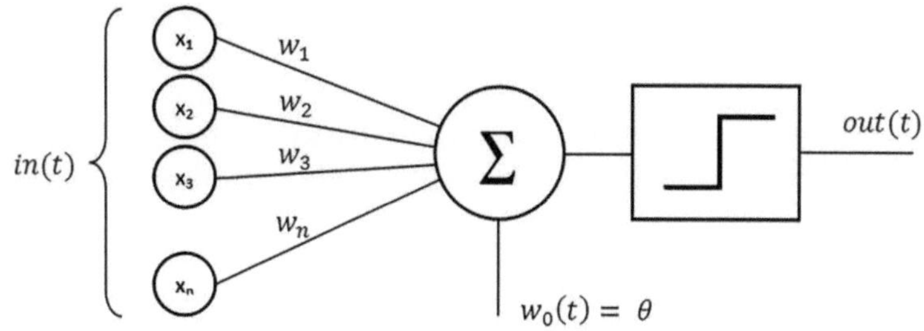


**Fig. 1.33** Diagram of a perceptron (https://commons.wikimedia.org/wiki/File:Perceptron_moj.png)

$$\begin{aligned} w_j &\leftarrow w_j + \eta\left(t^{(i)} - y^{(i)}\right)x_j^{(i)} \\ b &\leftarrow b + \eta\left(t^{(i)} - y^{(i)}\right) \end{aligned} \tag{1.32}$$

where η is the **learning rate**, a small positive constant controlling the step size of updates. Intuitively, if the perceptron predicts 0 but the true label is 1, it increases the weights of active input features, making it more likely to fire for similar examples in the future. Conversely, if it predicts 1 when the true label is 0, it decreases the weights of active inputs.

Rosenblatt proved the Perceptron convergence theorem, which states that if the training data are linearly separable, the algorithm is guaranteed to find a solution (a weight vector that classifies all examples correctly) in a finite number of steps. This result provided the first rigorous mathematical guarantee for learning in neural networks.

Despite its theoretical appeal, the perceptron has a fundamental limitation: it can only represent **linearly separable functions**. That is, it can only learn decision boundaries that are straight lines (in 2D), planes (in 3D), or hyperplanes (in higher dimensions).

The canonical counterexample is the XOR **(exclusive OR)** function. Consider two binary inputs $x_1$ and $x_2$. The XOR output is 1 if *exactly one* input is 1, and 0 otherwise. The positive points (1, 0) and (0, 1) are diagonally opposed, and the negative points (0, 0) and (1, 1) are the other diagonal. No single straight line can separate the positive from the negative points [137–139]. A perceptron trained on XOR will never converge to a correct solution. This limitation is not just an inconvenience—it defines the **expressive boundary** of the perceptron model.

In 1969, Marvin Minsky and Seymour Papert published the book Perceptrons, which analyzed perceptrons in depth. They proved mathematically that single-layer perceptrons cannot compute certain simple functions (like XOR) and argued that this made them fundamentally limited. While they acknowledged that networks with multiple layers could in principle overcome these limitations, no effective method for training multilayer networks was known at the time. Consequently, the field largely abandoned neural networks. Funding dried up, and AI research pivoted toward symbolic approaches and expert systems. This period of diminished interest and support became known as the AI winter.

In retrospect, the perceptron's failure was not conceptual but algorithmic. It lacked a mechanism to **propagate error signals backward** from the output to earlier layers. Without this, multilayer networks could not learn. But this insight would only be fully realized in the 1980s with the development of backpropagation, marking the rebirth of neural networks.

Despite its simplicity and historical setbacks, the perceptron remains a cornerstone of machine learning theory. It introduced fundamental ideas that persist in modern neural networks:

- **Weighted summation of inputs**—still the basic operation of every neuron.
- **Bias terms**—enabling flexible decision boundaries.
- **Activation functions**—now generalized beyond step functions to smooth non-linearities.
- **Iterative weight updates**—the foundation of all gradient-based learning methods.
- **Geometric interpretation**—the notion of learning as finding decision boundaries in feature space.

Even today, the perceptron is often the **first neural model taught** to students, because it encapsulates the essence of a neuron in its purest form. Understanding its structure and limitations sets the stage for appreciating the innovations of the multilayer perceptron.

The path from perceptron to multilayer perceptron reflects the central realization that **depth brings representational power**. By stacking multiple layers of perceptrons, each layer can transform the data into progressively more abstract features, enabling the network to solve non-linear problems [140–144]. But depth alone is insufficient without a learning mechanism. The next wave of progress came when researchers discovered how to **compute gradients through multiple layers** using the Backpropagation algorithm, coupled with the Gradient descent optimization method. This breakthrough would transform neural networks from theoretical curiosities into practical tools—and the multilayer perceptron would emerge as the first fully functional deep learning model.

While the perceptron marked a major conceptual milestone, it soon revealed a critical limitation: it can only solve **linearly separable problems**. This means that if the data points of two classes can be separated by a single straight line (in 2D) or a hyperplane (in higher dimensions), a perceptron can classify them. However, many real-world problems are **not linearly separable**.

A classic illustration is the Exclusive OR (XOR) problem. When plotted on a 2D plane, the XOR dataset forms two diagonal pairs of points belonging to two different classes. No straight line can separate these classes. This exposed a significant weakness: a single-layer perceptron simply cannot solve XOR.

This limitation stalled progress in neural networks for over a decade, especially after Marvin Minsky and Seymour Papert highlighted it in their 1969 book *Perceptrons*. It was not until the 1980s that researchers developed methods to train

**networks of multiple layers**—enabling them to learn complex, non-linear decision boundaries [145]. This gave rise to the **multilayer perceptron (MLP)**. A multilayer perceptron is essentially a **stack of perceptrons** arranged in layers. It has:

- an **input layer** that receives raw data features,
- one or more **hidden layers** where intermediate computations happen, and
- an **output layer** that produces the final prediction.

Each neuron in a layer is connected to every neuron in the subsequent layer (a **fully connected architecture**). Instead of binary step activation, MLP neurons typically use **differentiable activation functions** such as the Sigmoid function, Hyperbolic tangent (tanh), or Rectified Linear Unit (ReLU). These nonlinear functions are crucial: they enable the network to model nonlinear relationships (Fig. 1.34).

In an MLP, data flows **forward** from the input to the output layer, passing through the hidden layers. Each neuron computes a weighted sum of its inputs, adds a bias, applies an activation function, and passes the result forward. The network's overall behavior emerges from the collective transformation of data through all these layers [16, 146, 147].

The breakthrough that enabled MLPs to become practical was the Backpropagation algorithm. Introduced to the broader research community in the 1980s, backpropagation allowed neural networks to **learn internal weights efficiently** through gradient-based optimization. Training works as follows:

- The network produces predictions through **forward propagation**.
- The error between predictions and true labels is computed via a **loss function** (such as Mean Squared Error or Cross-Entropy Loss).
- This error is then **propagated backward** through the network using the chain rule of Calculus to compute gradients of the loss with respect to each weight.
- The weights are **updated using** Gradient Descent (or its variants), nudging them in the direction that reduces the loss (Fig. 1.35).

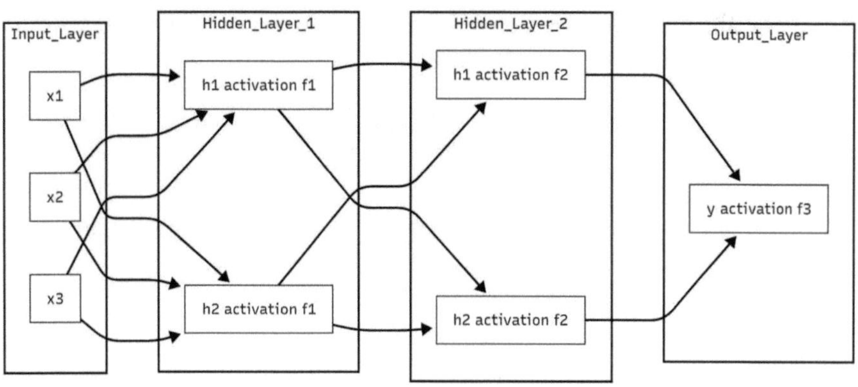


**Fig. 1.34** A schematic diagram showing a 3-layer MLP

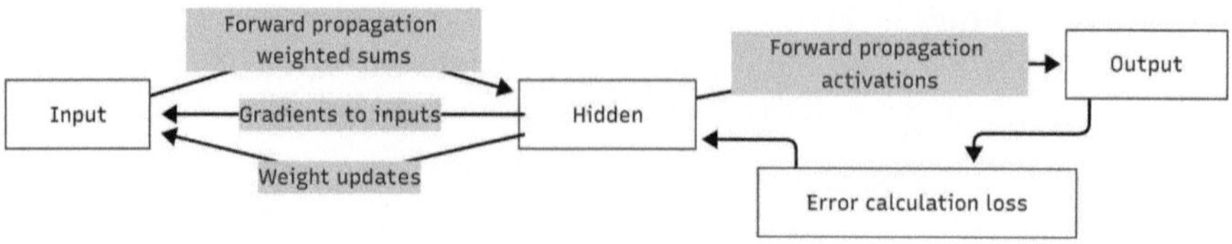


**Fig. 1.35** A flow diagram showing forward propagation

Backpropagation unlocked the power of MLPs, allowing them to learn highly nonlinear decision boundaries. As a result, MLPs became the foundation for many tasks such as speech recognition, character recognition, and early computer vision systems. At the heart of the Perceptron was a binary step function, simple, but non-differentiable and incapable of modeling smooth decision boundaries. The leap to powerful Multilayer Perceptron models required **differentiable and nonlinear activation functions**. These allow networks to approximate highly complex functions, which is essential because a purely linear stack of layers would collapse into a single linear transformation, no matter how deep [148]. Several activation functions became central:

- The **Sigmoid function** squashes real-valued inputs into the (0,1) interval, historically used in early MLPs.
- The **Hyperbolic tangent (tanh)** does something similar but outputs values in (−1,1), often resulting in faster convergence.
- The **Rectified Linear Unit (ReLU)** revolutionized deep learning by passing positive values unchanged while zeroing negatives. Its simplicity and ability to mitigate vanishing gradients made it the default choice in modern networks.

Choosing an activation function directly shapes how an MLP learns. Saturating nonlinearities like sigmoid can suffer from Vanishing gradient, while unbounded ones like ReLU enable deeper networks to train effectively. This insight paved the way for training networks with many hidden layers. An MLP can be formally described using linear algebra. Consider layer l with weight matrix $w^{(l)}$, bias vector $b^{(l)}$, and activation function *f*. Given input vector *x*, the forward pass is:

$$z^{(l)} = W^{(l)}a^{(l-1)} + b^{(l)}\ and\ a^{(l)} = f\left(z^{(l)}\right) \tag{1.33}$$

where $a^{(0)} = x$ and the final $a^{(L)}$ is the network output.

During backpropagation, we compute the gradient of the Loss function L with respect to each weight using the chain rule:

$$\frac{\partial \mathcal{L}}{\partial W^{(l)}} = \delta^{(l)}\left(\alpha^{(l-1)}\right)^T \tag{1.34}$$

where $\delta^{(l)}$ represents the error signal at layer l, computed recursively from the output layer backwards:

$$\delta^{(l)} = \left( W^{(l+1)^T} \delta^{(l+1)} \right) \Theta f'\left( z^{(l)} \right) \tag{1.35}$$

This machinery lets the network **learn by gradient descent**:

$$W^{(l)} \leftarrow W^{(l)} - \eta \frac{\partial \mathcal{L}}{\partial W^{(l)}} \tag{1.36}$$

where η is the Learning rate.

Building and training an MLP involves several subtle choices that greatly influence its success:

- **Weight initialization**: Poor initialization can stall learning. Modern practice uses strategies like Xavier initialization or He initialization to preserve variance of signals across layers.
- **Regularization**: Overfitting is common in large MLPs. Techniques like L2 regularization, Dropout, and Early stopping help maintain generalization.
- **Optimization algorithms**: Basic Stochastic gradient descent can be slow or unstable. Variants such as Adam or RMSProp adapt learning rates dynamically and often converge faster.
- **Data preprocessing**: Scaling inputs (normalization or standardization) ensures stable gradients and smoother convergence.

A major reason MLPs became a cornerstone of modern AI is their **theoretical universality**. The **Universal approximation theorem** states that a feedforward neural network with just **one hidden layer** and a sufficient number of neurons can approximate **any continuous function** on a compact domain, given appropriate weights and nonlinear activation functions.

This does **not** mean that one hidden layer is always optimal in practice — it only guarantees that such a solution exists. In reality, deeper architectures (more layers with fewer neurons each) often learn more efficiently and generalize better, especially for high-dimensional and hierarchical data.

The universal approximation result placed MLPs alongside classical function approximators (such as Polynomial regression or Fourier series) but with the crucial advantage that their features are **learned automatically** from data rather than handcrafted. This insight marked a paradigm shift in machine learning: representation learning could be done **end-to-end** through gradient-based training. Before the rise of deep convolutional and transformer architectures, MLPs dominated many AI tasks [149]. They remain widely used today, especially where data is **tabular or low-dimensional** and **interpretability and training efficiency** are priorities.

**Classic applications include:**

- **Pattern recognition**—early systems for handwritten digit classification (such as MNIST) relied on MLPs.
- **Speech and audio processing**—phoneme classification and isolated word recognition.

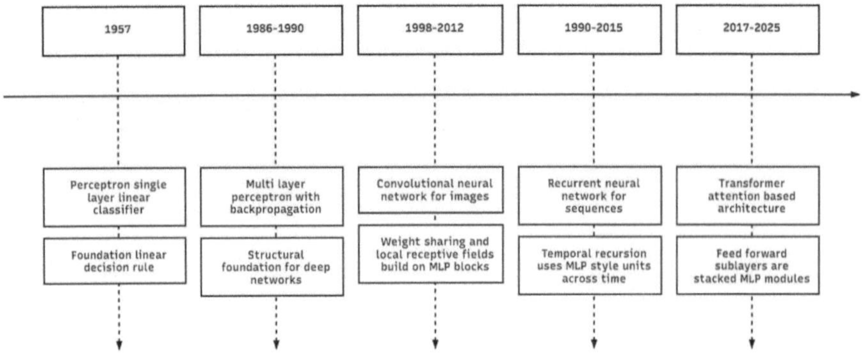


**Fig. 1.36** A timeline graphic showing perceptron evolution

- **Time-series forecasting**—stock price prediction, weather modeling, and demand forecasting.
- **Medical diagnostics**—classifying patient records or lab results based on structured features.
- **Industrial control systems**—modeling sensor-response relationships and controlling nonlinear processes.

Even today, MLPs are frequently used as **baseline models** for new datasets and as **components within larger architectures**, such as Multilayer perceptron mixer (**MLP-Mixer**) blocks in computer vision models or the feedforward layers inside Transformer networks. Despite their flexibility, MLPs have some inherent limitations:

- They do not exploit **spatial locality** (as Convolutional neural networks do) or **temporal sequence structure** (as Recurrent neural networks do).
- They can require large amounts of labeled data to generalize well.
- Their dense connectivity makes them **parameter-heavy** and sometimes computationally inefficient.

These challenges spurred the development of specialized architectures. Yet, understanding MLPs remains essential because they form the **conceptual blueprint** for nearly all modern neural networks: layered structures, differentiable transformations, and gradient-based learning (Fig. 1.36).

## 1.7 Fundamentals of Neural Networks

**Neural networks (NNs)** are computational models of neuron networks in the human brain. The neural network is a collection of very simple units (neurons), which are organized in layers and are interconnected in a network by employing weighted links. The neuron, in order to produce an output, calculates its weighted sum of inputs and uses an activation function. With the previously described basics, neural networks learn complex functional correlations from input by changing connection weights.

The concept of NNs dates back to the 1940s when McCulloch and Pitts proposed a simple binary neuron model. In 1958, Frank Rosenblatt introduced the perceptron, an early neural network with a single trainable layer, demonstrating a machine's ability to learn from data. These early networks, which were limited (a single-layer perceptron can only learn linearly separable patterns), were met with skepticism following Minsky and Papert's 1969 examination of their limits. Neural network development halted during the 1970s, a period known as the "AI winter". Interest in the field was revived in the 1980s with the development of multi-layer networks and the introduction of the backpropagation training algorithm. Inspiring synthesis and correlation of various branches of understanding brought the essentials of modern NNs. Modern deep neural networks (with many layers) have revolutionized machine learning by automatically learning multi-level data representations [150]. They currently achieve human-level performance in a variety of tasks, including picture and speech recognition. In the following sections, we will go over the fundamental concepts of neural networks, such as common architectures, loss functions, and training optimization algorithms, regularization methods to prevent overfitting, training challenges (such as convergence difficulties and vanishing/exploding gradients), and training stabilization methods such as batch normalization.

### *1.7.1 Neural Network Architectures*

Neural network architecture specifies how neurons are connected and structured. Two main categories of topologies are **feedforward networks and recurrent networks**, which vary in how information is passed through them. The goal of a neural network is to learn a function: f(x;θ) that maps input vectors x to output vectors, where θ represents all trainable parameters such as weights and biases. This mapping is achieved by layering countless layers of computation, each of which is comprised of linear transformations in sequence by nonlinear activation functions. The most remarkable contributions to neural network architecture have been the introduction of efficient activation functions. Due to its ease of use and performance, the Rectified Linear Unit (ReLU) has become the de facto standard for neural network hidden layers [151, 152]. The mathematical definition is: ReLU(z) = max (0, z), which prevents the vanishing gradient problem and accelerates convergence during training (Fig. 1.37).

**Feedforward neural networks**, also known as **multi-layer perceptrons (MLPs)** or **deep feedforward networks**, are the most fundamental neural network architecture. A feedforward network transmits information in a single path from the input layer to the output layer, with no cycles or feedback links. In a more simplified conceptualization, they are computationally straightforward and concise [153].

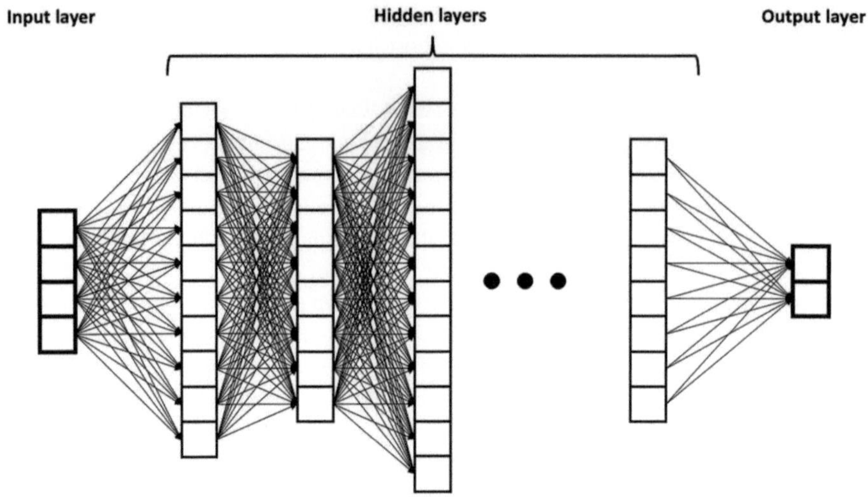


**Fig. 1.37** A neural network (https://commons.wikimedia.org/wiki/File:Example_of_a_deep_neural_network.png)

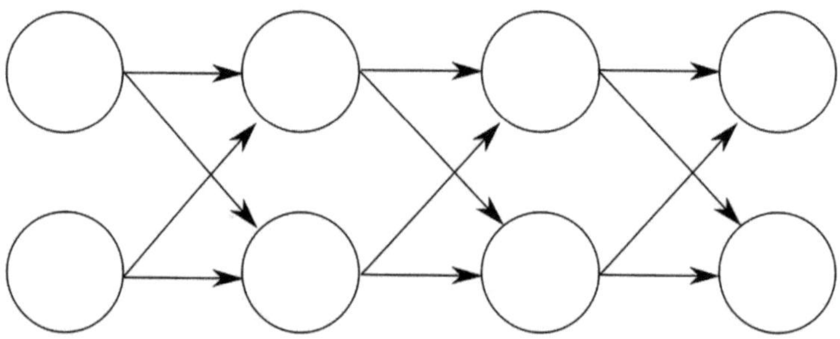

**Fig. 1.38** Feedforward neural network (https://commons.wikimedia.org/wiki/File:Feedforward_network.svg)

Each neuron in a layer takes input from the previous layer, and its result is transmitted to neurons in a succeeding layer (Fig. 1.38).

Mathematically, a feedforward network computes a function y = f(x), through a sequence of transformations layer-wise. For an L-layer network (including only hidden and output layers, excluding the input), backpropagation function are given below (Table 1.13).

These activation functions introduce a level of nonlinearity, allowing the network to learn intricate patterns. In current neural networks, ReLU is usually the default activation because of its characteristics in reducing the vanishing gradient problem and boosting trainability. For example, using ReLU in a hidden layer, the transformation is given by:

**Table 1.13** Hidden Layer and Activation Functions

| Component | Description | Functions |
|---|---|---|
| Layer 1 (hidden layer 1) | First hidden layer of the network. | $h^{(1)} = g^{(1)}(W^{(1)}x + b^{(1)})$ |
| Layer 2 (hidden layer 2) | Second hidden layer of the network. | $h^{(2)} = g^{(2)}(W^{(2)}h^{(1)} + b^{(2)})$ |
| Layer L (output layer) | Final layer producing the network's output. | $y = g^{(L)}(W^{(L)}h^{(L-1)} + b^{(L)})$ |
| **Activation functions** | Nonlinear functions used to introduce nonlinearity into the model: | |
| Sigmoid function | Maps input to a range between 0 and 1. | $\sigma(z) = \frac{1}{1+e^{-z}}$ |
| Hyperbolic tangent (tanh) | Maps input to a range between −1 and 1. | $\tanh(z) = \frac{e^{2z}-1}{e^{2z}+1}$ |
| Rectified linear unit (ReLU) | Outputs zero for negative inputs and the input itself for positive values. | (ReLU) : $r(z) = max\,(0, z)$ |

$$h(l) = \max\left(0, W^{(l)}h^{(l-1)} + b^{(l)}\right), \tag{1.37}$$

applied elementwise to the preactivations:

$$z^{(l)} = W^{(l)}h^{(l-1)} + b^{(l)} \tag{1.38}$$

The output layer of a feedforward net is specified based on the specific problem. For binary classification, a softmax output layer is typically used to produce a probability P (y = 1|x) of the positive class. For multiclass classification, a softmax output layer is used to output a probability distribution over d classes. If in = ($in_1$, …, $in_d$) represents the input scores to the softmax layer (typically described as logits), then softmax produces a d-dimensional vector where the k-th component is defined as:

$$softmax(in)k = \frac{e^{in_k}}{\sum_{k=1}^{d} e^{in_k}} \tag{1.39}$$

for k = 1, …, d. This guarantees that there are output values in the range summing to 1, and thus are [0,1] interpretable as probabilities of a class.

**Softmax** outputs, by definition, sum to 1 and are non-negative, and so may be thought of as probabilities of classes. For regression (prediction of output as a continuous value), a linear output neuron is most commonly employed (no activation function), outputting directly as a real value (or a few such linear output neurons for multi-output regression). The design of the output layer and activation function thus depends on whether the network is classifying or regressing, and is related to the choice of loss function.

Almost all feedforward networks are densely connected, i.e., a neuron in a layer is connected to every other neuron in the next layer. There are, however, several specialized feedforward architectures in place to capitalize on structure in data. The ultimate classic example is that of a convolutional neural network (CNN), where a grid-based receptive field and weight sharing are imposed, significantly decreasing the number of parameters and increasing efficiency for image and signal data. The CNNs are, in fact, feedforward networks of limited connectivity and have been very successful in computer vision [154]. Regardless of these departures, what characterizes a feedforward network is the acyclic flow of information from inputs to outputs.

While feedforward networks work in input/output sets of a fixed size, most problems consist of sequential data of a variable length (such as time series and sentences), the **recurrent neural network** is a network structure that operates on sequential data by employing feedback links, which enable information to persist over multiple time steps. The neurons in an RNN have directed cycles, from which an internal state (memory), which holds prior context, arises [155]. That is why a network is in a position to learn temporal dynamics and sequence dependencies beyond feedforward models' abilities.

An RNN is a sequence of identical modules, each for a time step, passing information to the next. Consider providing a sequence of input to an RNN as a means of generating $\{y_1, y_2, \ldots, y_T\}$ output. The RNN has a hidden state at time t computed from $\{x_1, x_2, \ldots, x_T\}$. We may define a simple ("vanilla") RNN by the recurrence:

$$\begin{aligned} h_t &= f\left(W_{xh} x_t + W_{hh} h_t - 1 + b_h\right) \\ y_t &= g\left(W_{hy} h_t + b_y\right) \end{aligned} \tag{1.40}$$

where $W_{xh}$, $W_{hh}$, and $W_{hy}$ are input-to-hidden, hidden-to-hidden (recurrent), and hidden-to-output connection weights, respectively, and are activation functions $f(\cdot)$, $g(\cdot)$ (Fig. 1.39).

Depending upon how it is to be applied, recurrent networks may be one-to-one (basic networks), one-to-many (e.g., image captioning, where a sequence output is derived from a single input image), many-to-one (e.g., sequence classification, sentiment analysis, where a whole sequence produces a single output), or many-to-many (e.g., sequence-to-sequence models for machine translation, where an input sequence is converted to an output sequence). Standard RNNs, as mentioned previously, suffer from being unable to learn long-range relationships in sequences because of problems relating to **vanishing and exploding gradients**. More advanced RNN architectures have been designed to address this, foremost among which is the **Long Short-Term Memory (LSTM)** network and **Gated Recurrent Unit (GRU).** The LSTM carries gating methods that regulate information flow, so that, during its processing, it is able to preserve long-term information and minimize difficulties relating to vanishing gradients. Indeed, LSTMs, as proposed by Hochreiter and Schmidhuber in 1997, and GRUs have pretty much supplanted vanilla RNNs for learning from elongated sequences, as a result of their capacity to pass along information over dozens or hundreds of time steps. Even so, recurrent

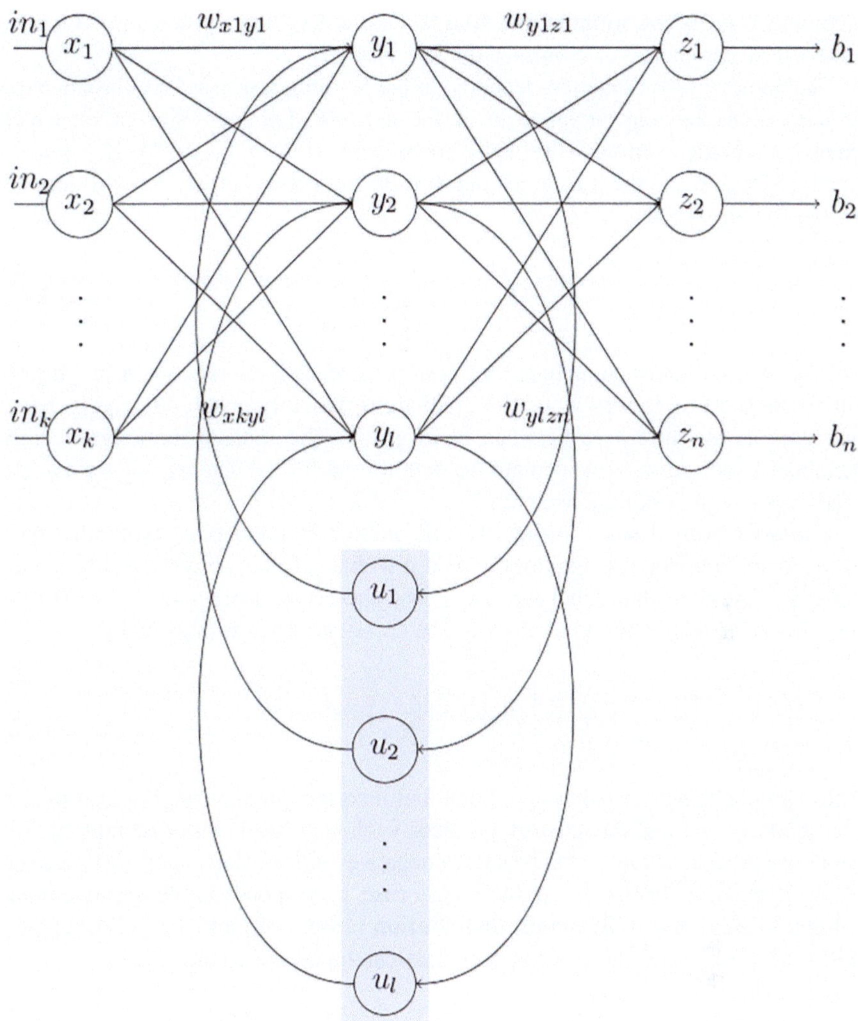


**Fig. 1.39** A simple recurrent neural network (https://commons.wikimedia.org/wiki/File:Elman_srnn.png)

networks are computationally expensive for longer sequences (owing to sequential processing). Modern sequence modeling architecture, such as the Transformer (which applies only attention mechanisms, not recurrent links), has increasingly become popular, though it is outside our scope as an article concerning foundational neural network architecture.

To train a neural network, it is important to assess its performance on the training data. A **loss function,** also known as a cost function or objective function, helps quantify this performance. The loss function measures the difference between the network's predictions L(θ) and the actual target values based on the current parameters, including **weights and biases.** Optimizing θ is a process in which we adjust these parameters to minimize the loss [156].

The choice of loss function depends on the learning task (such as classification or regression) and the output layer of the network. For regression tasks, which involve predicting continuous values, a common loss function is the **Mean Squared Error (MSE)**. If y is the true target and $\hat{y}$ is the network's output, the MSE loss for a single example is defined as:

$$L_{MSE} = \frac{1}{2}(y - \hat{y})^2 \tag{1.41}$$

The factor of 1/2 is often included for convenience, as it cancels out during differentiation. The loss is zero when the predictions perfectly match the targets, but it increases as the difference between them grows. The squared term significantly penalizes larger errors. To calculate the overall loss for the dataset, we average the MSE loss over all training instances.

**Cross-Entropy Loss** is primarily utilized for classification, especially with probabilistic outputs. It is commonly paired with a softmax output layer. In binary classification, the actual label y ϵ {0,1}, and the network prediction $\hat{y}$ ϵ [0,1] is interpreted as the likelihood of class 1. The cross-entropy loss is given by:

$$L_{CE} = -\left[y \log \hat{y} + (1 - y) \log\left(1 - \hat{y}\right)\right] \tag{1.42}$$

This formula heavily penalizes confident but incorrect predictions. For example, if the true label is 1 and the network predicts $\hat{y}$ close to 0, the loss becomes significantly large because log $\hat{y}$ will be a large negative number. Cross-entropy is derived from information theory and, in the binary case, corresponds to the negative logarithmic likelihood of a **Bernoulli distribution** [154]. For multiclass classification with softmax outputs, the cross-entropy loss can be expressed as:

$$L = -\sum_{k=1}^{d} y_k \log \hat{y} \hat{k} \tag{1.43}$$

Here, $y_k$ is 1 for the correct class and 0 for others (one-hot encoding), while $\hat{y}\hat{k}$ represents the predicted probability for class k.

In addition to the commonly used loss functions, other types may be beneficial in specific situations. For instance, **L1 loss** (which uses absolute differences), **Huber loss** (which performs better in the presence of outliers), and **reconstruction loss** (used in unsupervised models such as autoencoders) are important depending on the type of task and data.

**Gradient descent,** a first-order optimization algorithm, is the primary technique used for training neural networks. This algorithm computes the gradient (the vector of partial derivatives) of the loss function with respect to all parameters, $\Delta_\theta L$. It involves updating the parameters in the opposite direction to the gradient, which is the direction of the steepest descent on the loss surface. A simple update for a weight parameter w in the network is:

$$w \leftarrow w - \eta \frac{\partial \mathcal{L}}{\partial w} \tag{1.44}$$

where $\eta$ is the learning rate, a small positive scalar that regulates the step size. By iteratively applying this update to every parameter, the loss gradually decreases, and the network's predictions ideally become more accurate.

A challenge arises because $\nabla_\theta L$ can be computationally expensive if done naively, as it depends on all layers of the network. This is where the backpropagation algorithm comes into play. **Backpropagation** is an efficient method for computing gradients in a neural network by propagating error signals backward from the output layer to the input layer, utilizing the chain rule of calculus. This method enables the calculation of the gradient of the loss with respect to each weight systematically, layer by layer [157, 158]. The development of backpropagation was a significant advancement, enabling effective training of multi-layer networks.

While plain SGD is effective, it can be slow to converge or may get stuck in certain regions of the loss surface. Over the years, several enhancements to SGD have been developed to accelerate training and improve robustness [159]. These include:

- Momentum: SGD with momentum accumulates an exponential moving average of past gradients to smooth out updates.
- Adaptive learning rates: Methods like AdaGrad, RMSProp, AdaDelta, and Adam automatically adjust the learning rate for each parameter based on the history of past updates. This approach enables the network to learn more efficiently, particularly when certain parameters necessitate larger or smaller updates. For instance, AdaGrad reduces the learning rate for parameters with significant early gradients, while RMSProp utilizes a moving average of squared gradients to normalize learning rate updates.
- Adam optimizer: Adam is one of the most popular optimizers because it combines momentum with adaptive learning rates. It remembers previous gradients to smooth out updates and adjusts updates based on the magnitude of earlier gradients. Adam also tracks the average of past gradients (first moment $m_t$) and the average of squared gradients (second moment $v_t$).

**Regularization** is an important technique used in deep learning to reduce overfitting, which is the tendency of a model to memorize the training data and perform poorly on unseen data. Overfitting occurs when a model learns not only the true underlying patterns in the training data but also its noise, leading to poor performance on new inputs. To improve a model's ability to generalize to new data, regularization modifies the learning process to limit the model's complexity. A more complex model might match the training data well but still fail to generalize effectively. By introducing additional data or constraints, regularization techniques enable the model to choose simpler solutions that generalize better. In neural networks, two of the most common regularization methods are L2 weight regularization (also known as weight decay) and dropout.

**L2 regularization** works by adding a penalty term to the loss function, which corresponds to the square of the weights' magnitudes. This modification encourages smaller weights, thereby promoting a simpler model that can generalize more effectively. Dropout, on the other hand, is a technique that improves generalization by randomly deactivating a portion of neurons during each training step. This prevents the model from assigning excessively large weights to specific neurons, resulting in a smoother and more straightforward model. Essentially, dropout simulates the training of an ensemble of smaller subnetworks, forcing the model to learn more robust features that are not reliant on specific neurons [160, 161]. Regularization modifies the loss function that the model aims to minimize during training. By adding a penalty term that reflects the model's complexity (like weight size in L1 or L2 regularization), the training process is guided toward solutions that balance simplicity and performance.

The standard loss function, as shown in linear regression, is the mean squared error (MSE), represented in the following way:

$$\min_{w^{(i)}}\left[\frac{1}{N}\times\sum_{i=1}^{N}\left(f\left(x_i\right)-y\right)^2\right] \tag{1.45}$$

In this equation, the goal is to minimize the difference between the prediction f(x) and the observed value y. Here, f(x) represents the regression line, which can be expressed as:

$$f = w^{(i)}x^{(i)} + b \tag{1.46}$$

Consequently, the algorithm aims to minimize the MSE to determine the values of the parameters w and b from the training set. If any of the parameters in the model are close to zero, the model will be less complex.

**Weight decay,** also known as **L2 regularization or Ridge Regression**, has long been a standard method for improving the generalization performance of neural networks. It penalizes large weights in the cost function, which helps to discourage high weight values and steer the network towards simpler models that are less prone to overfitting.

L2 regularization increases the regularization term by taking the square of the weights. As a result, this approach produces models with more evenly distributed weights and typically fewer variables compared to L1 regularization [163, 164]. While L1 regularization tends to promote sparsity by zeroing out some weights, L2 regularization spreads the weight across all variables, making it particularly useful in scenarios involving a large number of highly correlated variables.

The initial equation is modified to integrate L2 regularization as follows:

$$\min_{w^{(i)}}\left[c\times\sum_{i=1}^{N}\left(w^{(i)}\right)^{2}+\frac{1}{N}\times\sum_{i=1}^{N}\left(f\left(x_{i}\right)-y\right)^{2}\right] \tag{1.47}$$

By reducing the impact of noise or outliers on the variables, L2 regularization can enhance model stability, especially when the training data is limited. It adds a term proportional to the square of the weight coefficients to the cost function:

$$J\left(w_{1},w_{2},b\right)=\frac{1}{n}\sum_{i=1}^{n}\left(y_{i}-\widehat{y}_{i}\right)^{2}+\lambda\left(w_{1}^{2}+w_{2}^{2}\right) \tag{1.48}$$

Although L2 regularization does not zero out any weight coefficients like L1 regularization, it tends to reduce the magnitude of all the weight coefficients, leading to a more balanced model.

**Dropout** is a widely used and effective regularization technique in deep learning. Its main objective is to reduce overfitting by preventing neural networks from becoming overly reliant on specific neurons. This is achieved by randomly "dropping out" a subset of neurons during each iteration of the training phase. During training, each neuron is kept with a certain probability p (commonly set at 0.5 for hidden layers), while with a probability of 1-p, it is set to zero. The deactivated neurons do not participate in the forward pass or in the backpropagation step. This process results in slightly different network architectures at each iteration, effectively training an ensemble of numerous smaller subnetworks.

To ensure consistency between training and inference, all neurons are active during testing time, but their outputs are scaled by the same dropout probability p. Some implementations manage this scaling during training, eliminating the need for adjustments at inference time [162]. Dropout helps to disrupt complex co-adaptations among neurons. Since any neuron can be randomly dropped, the network is encouraged to learn robust, independent features that remain useful across diverse combinations. This significantly enhances the network's ability to generalize to new, unseen data.

Another perspective on dropout is that it enables the effective training of a large network of neural networks. If a network contains n neurons, then deleting various combinations can result in many subnetworks. In practice, dropout allows the model to train and sample from this vast ensemble without requiring a substantial amount of computing power. Dropout offers several benefits, including improved generalization (by significantly reducing overfitting and making the network less sensitive to noise in the training data), model robustness (encouraging neurons to be useful on their own and in various random combinations), flexibility (being easy to implement and compatible with most deep learning architectures), and versatility (applicable to fully connected layers, convolutional layers, and even recurrent neural networks with proper adaptation). However, dropout does lengthen training time, as the added randomness can slow down convergence. The dropout rate (p) is a hyperparameter that must be carefully calibrated based on the architecture and task, typically ranging from 10% to 50%. Usually, output neurons are not included in dropout to ensure consistent predictions.

### 1.7.2 Convergence and Vanishing/Exploding Gradients

In machine learning, **convergence** refers to the point at which a model's parameters and the value of its loss function stabilize and stop changing significantly with additional training. In deep neural networks, particularly multi-layered ones, this convergence can be hindered by vanishing and exploding gradients. A partial derivative, which is a derivative of a function with respect to a single variable (such as x or y), is used to compute gradients. As neural networks grow deeper, gradients calculated through backpropagation may either vanish (become extremely small) or explode (become extremely large). This phenomenon was first analyzed in the context of training recurrent networks but also applies to deep feedforward networks.

The **vanishing gradient** problem presents a substantial challenge for training deep neural networks. This issue occurs when the gradients of the loss function become exceedingly small as they propagate backward through the layers of the network during training. When the gradients are too small, the model learns very slowly or may not learn at all, as the updates to the weights become negligible. During backpropagation, gradients are computed using the chain rule, which involves multiplying the derivatives of activation functions layer by layer from the output layer back to the input layer. In complex networks, these gradients may decline exponentially due to this repeated multiplication [165, 166]. This is especially problematic when using activation functions like tanh or sigmoid, whose derivatives are less than 1 for most of their range. As a result, gradients that reach the earlier layers of the network may become so tiny that they effectively disappear, making it difficult for those layers to learn meaningful representations.

As a consequence of the vanishing gradient problem, the early layers of the network learn very slowly, if at all. These layers are responsible for detecting fundamental patterns in input data (such as edges in image processing). If they do not receive adequate training, the overall performance of the network suffers, leading to poor convergence, prolonged training times, and suboptimal accuracy. Several strategies have been developed to mitigate the vanishing gradient problem:

- Popular activation functions such as Rectified Linear Units (ReLUs) and their variants (like Leaky ReLU and ELU) promote healthy gradient flow by avoiding saturation in the positive range.
- Batch normalization helps reduce the likelihood of extreme values and facilitates training by normalizing activations across mini-batches.
- Techniques for meticulous weight initialization, such as He initialization and Xavier (Glorot) initialization, help preserve the gradient scale.

- Residual connections, originally used in ResNet architectures, provide shortcut pathways that allow gradients to traverse the network more directly, bypassing certain layers.

Another significant challenge in training deep neural networks is the **exploding gradient** problem, which is essentially the opposite of the vanishing gradient problem. In this case, gradients become excessively large during backpropagation, resulting in unstable weight updates. This may cause the training process to diverge rather than converge toward an optimal solution, leading to a rapid and erratic increase in the model's loss function.

Similar to the vanishing gradient issue, the exploding gradient problem occurs during the backpropagation process. While calculating gradients through the layers of a deep neural network, the chain rule is used to multiply derivatives from one layer to the next. If the weights and derivatives exceed one, gradients may increase exponentially as they propagate backward through the network. This behavior is more likely in very deep networks containing many layers. The signal is amplified when the initial weights are set too high, and under certain conditions, some activation functions further enhance gradients. When gradients become excessively large, the updates to the model's weights can grow so substantial that the model deviates from optimal values [167]. This can lead to numerical instability, causing the model's parameters to explode uncontrollably. In extreme situations, calculations may yield NaNs (Not-a-Number), making learning impossible, as the model fails to converge. Several approaches can be taken to address the exploding gradient problem:

- Gradient clipping is a common technique that limits gradients if they surpass a predefined threshold, preventing drastic updates that could destabilize training.
- Proper weight initialization techniques, such as Xavier or He initialization, can assign weight values that are less likely to lead to gradient explosions.
- Batch normalization and other normalization techniques can help stabilize the values moving through the network, reducing the likelihood of producing excessively large activations and gradients.

**Batch normalization (BN)** is a fundamental technique in deep learning that significantly enhances the efficiency and stability of neural network training. It addresses the problem of internal covariate shift, which occurs when changes in the parameters of previous layers affect the distribution of activations. This shift complicates training, often necessitating careful initialization to maintain stability and slower learning rates.

BN is an algorithmic strategy that accelerates and stabilizes the training of Deep Neural Networks (DNNs). It works by normalizing activation vectors from hidden layers using the mean and variance of the current batch. This normalization is applied before or immediately after the non-linear activation function. BN normalizes the activations of each layer in the neural network to have a mean of zero and a variance of one. This normalization is applied to batches, which are typically small samples of the dataset, with sizes of 32 or 64 randomly selected examples. BN utilizes the mean and variance of the activations for each batch to normalize the values before or sometimes after they pass through the non-linear activation functions.

While normalization of input data is often performed during preprocessing, it can be disrupted by changes introduced by activation functions in hidden layers. BN restores a consistent distribution and stabilizes the learning process by renormalizing values at each layer. This stability accelerates convergence, reduces reliance on careful weight initialization, and allows for higher learning rates [168]. Consequently, Batch Normalization enhances the model's overall robustness and speeds up training. It also mitigates common issues in deep networks, such as exploding or vanishing gradients.

Although specialized methods like forget gates in Long Short-Term Memory (LSTM) networks manage sequence-specific challenges, BN provides a generalized solution applicable to various network topologies, including feedforward, convolutional, and recurrent networks. Thus, Batch Normalization represents a significant advancement in deep learning. Maintaining consistent activation distributions throughout training improves model performance, training speed, and stability across an array of neural network models and tasks, such as image recognition and natural language processing.

The **normalization of inputs** for each mini-batch is the first step in BN. This process ensures that the activations have a mean of zero and a unit variance for each feature within a layer. For a mini-batch of size m, the activations can be represented as $B = \{x_1, x_2, \ldots, x_m\}$, where $x_i$ denotes the activation for the i-th instance in the batch. The mean $\mu_B$ and variance $\sigma_B^2$ for a feature across the mini-batch are calculated as follows:

$$\mu_B = \frac{1}{m}\sum_{i=1}^{m} x_i \tag{1.49}$$

$$\sigma_B^2 = \frac{1}{m}\sum_{i=1}^{m} (x_i - \mu_B)^2 \tag{1.50}$$

Each activation $x_i$ is then normalized using the computed mean and variance:

$$\hat{x}_i = \frac{x_i - \mu_B}{\sqrt{\sigma_B^2 + \epsilon}} \tag{1.51}$$

Here, $\epsilon$ is a small constant added to the variance to avoid division by zero, often referred to as a numerical stabilizer.

Even if the data has been normalized, the network may still benefit from adjusting these standardized values to better capture the underlying patterns. This adjustment is achieved through scaling and shifting.

During this step, two additional parameters, $\gamma$ (gamma, the scale factor) and $\beta$ (beta, the shift factor), are introduced to further transform the normalized activations ${}^{(i)}$. Similar to the network weights, these parameters are learned during training. The transformation can be defined as:

$$y_i = \gamma \hat{x}_i + \beta \tag{1.52}$$

In this equation, ${}^{(i)}$ denotes the output of the Batch Normalization layer, while $\gamma$ and $\beta$ are the parameters learned alongside the initial model parameters. Their significance lies in allowing the model to reverse the normalizing effect if it minimizes the loss effectively.

**Batch normalization** offers several key advantages that improve neural network training. First, by stabilizing input distributions between layers, it enables faster convergence, better learning rates, and reduces the number of training epochs required. Second, it enhances training stability by decreasing sensitivity to parameter value changes and maintaining consistent activation patterns, which mitigates internal covariate shift. Additionally, Batch Normalization eliminates the need for precise weight initialization. By rescaling activations, it circumvents problems such as vanishing or exploding gradients, thereby increasing training flexibility and reducing dependence on specific initialization strategies.

Neural networks are widely used across various industries due to their ability to learn complex patterns from data. Here are some key application areas [169]:

1. Computer vision: Convolutional Neural Networks (CNNs) excel in tasks such as image classification, object recognition, and medical diagnostics. The success of AlexNet on the ImageNet dataset marked a significant advancement in deep learning applications within the field of computer vision.
2. Natural language processing (NLP): Recurrent Neural Networks (RNNs) and Transformers facilitate machine translation (for example, Google Translate), speech recognition, and conversational AI. Models like BERT and GPT offer a profound understanding of language context.
3. Healthcare: Neural networks can predict patient risk, diagnose diseases through imaging, and tailor treatments to individual needs. Notably, deep learning models such as DeepMind's AlphaFold have successfully predicted protein structures.
4. Finance: In the finance sector, neural networks are employed for detecting fraud, assessing credit scores, and algorithmic trading. They do this by modeling complex transaction patterns and identifying anomalies in real time.
5. Autonomous systems: Autonomous vehicles utilize CNNs for visual perception and Long Short-Term Memory networks (LSTMs) for control and decision-making.

These applications demonstrate the versatility and potential of neural networks in processing and interpreting vast amounts of data across various fields.

## 1.8 Advanced Network Classifications

Neural networks have significantly advanced the field of machine learning by employing various complex methods to recognize patterns in data. Among these methods are convolutional neural networks (CNNs), recurrent neural networks (RNNs), long short-term memory networks (LSTMs), gated recurrent units (GRUs), transformer architectures, attention mechanisms, graph neural networks, and transfer learning with pretrained models.

**Convolutional neural networks** are among the most widely used neural networks in deep learning and have propelled machine learning to new heights, particularly in areas such as face recognition and autonomous vehicles [168]. Furthermore, graph neural networks are a class of models that learn by gathering information from neighboring nodes, which allows them to learn not only from data but also from each other.

Convolutional neural networks (CNNs) have emerged as a leading type of neural network in the realm of deep learning. They have played a crucial role in various advancements that were once considered impossible, including progress in facial recognition, autonomous vehicles, self-service in supermarkets, and breakthroughs in medical research. CNNs are a form of feedforward neural network designed to automatically extract features from data through convolutional structures, eliminating the need for manual feature extraction.

The architecture of CNNs is inspired by visual perception; each CNN kernel acts as a distinct receptor responding to different features. Activation functions simulate the mechanism whereby only electrical signals from neurons that surpass a specific threshold can be transmitted to the next neuron. To ensure that the CNN system learns what to expect, loss functions and optimizers were developed [170–174]. Compared to fully connected (FC) networks, CNNs offer several advantages:

1. Local connections: Each neuron is connected only to a limited number of neurons from the previous layer rather than every neuron, which effectively reduces parameters and speeds up convergence.
2. Weight sharing: Weights can be shared among a group of connections, further decreasing the number of parameters.
3. Downsampling (dimension reduction): This technique helps reduce the amount of data while preserving important information by using pooling layers that capitalize on the principles of local correlation in images, which also aids in eliminating redundant features (Fig. 1.40).

To build a CNN model, four key components are essential. Convolution is a critical step in feature extraction, resulting in outputs referred to as feature maps. When a convolution kernel of a specific size is set up, some information is lost at the borders; thus, padding is introduced to enlarge the input with zeros, allowing for indirect size adjustments. Additionally, to manage the diversity of convolution, the stride is employed: a larger stride results in a lower density of features, and a smaller stride yields a denser output [175].

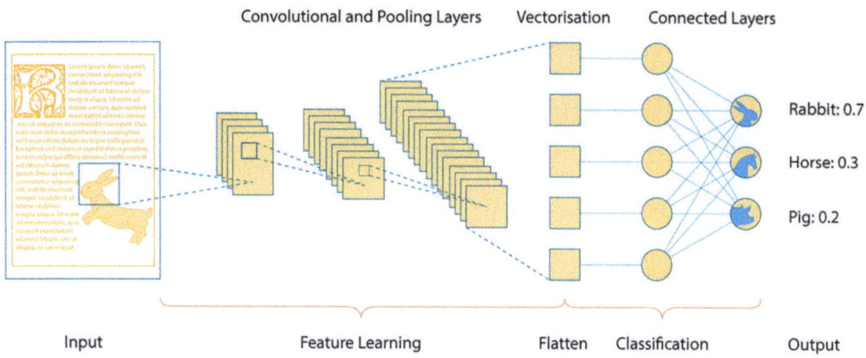


**Fig. 1.40** Diagram of CNN (https://commons.wikimedia.org/wiki/File:Convolutional_Neural_Network.png)

After the convolution process, feature maps contain a vast number of features, which may lead to overfitting. To counteract this, downsampling (or pooling) methods, such as max pooling and average pooling, are utilized to prevent redundancy. Moreover, to enable convolution kernels to perceive larger areas, dilated convolution was proposed. This technique employs a standard 3x3 convolution kernel, alongside a two-dilated 3x3 kernel and a four-dilated 3x3 kernel. It's important to note that there are zero-valued gaps introduced between each point of the convolution kernel [176, 177]. Although the two-dilated convolution achieves a 7x7 receptive field and the four-dilated version attains a 15x15 receptive field, the valid kernel points remain at 3x3.

Typically, a CNN processes an order-3 tensor as its input, such as an image with H rows, W columns, and 3 color channels (R, G, B). CNNs can also handle higher-order tensor inputs in a similar manner. The input undergoes a series of processing steps through multiple layers, which may include convolution layers, pooling layers, normalization layers, fully connected layers, and loss layers. The abstract representation of a CNN illustrates this process:

$$x^1 \rightarrow w^1 \rightarrow x^2 \rightarrow \cdots \rightarrow x^{L-1} \rightarrow w^{L-1} \rightarrow x^L \rightarrow w^L \rightarrow z \tag{1.53}$$

The equation above outlines the layer-by-layer operation of a CNN during a forward pass.

Despite the many advantages that CNNs offer, they also come with some disadvantages. These include the loss of spatial information (which can result in important details being overlooked during processing), high computational costs (as achieving high accuracy can be resource-intensive), challenges with deployment (difficulties in real-time deployment due to their size and complexity), and vulnerability to data poisoning (where CNNs can be misled during training), among others [178].

**Recurrent Neural Networks (RNNs)** are a type of supervised machine learning model characterized by one or more feedback loops, contrasting with the feedforward architecture of CNNs. These feedback loops create recurrent cycles over a specific time or sequence. To train an RNN in a supervised manner, a dataset of input-target pairs is required. The primary goal is to minimize the difference between the output and the target pairs (i.e., the loss value) by optimizing the network's weights.

A simple RNN comprises three layers: the input layer, the recurrent hidden layer, and the output layer. The input layer consists of N input units, which receive sequences of vectors over time t. In a fully connected RNN, these inputs are connected to the hidden units in the hidden layer, where weights determined by a weight matrix define the connections. The hidden layer contains M hidden units ($h_t$), which are interconnected through recurrent connections [179, 180]. The equation that describes the state space or "memory" of the system in the hidden layer is as follows:

$$h_t = f_H(o_t), \tag{1.54}$$

where

$$o_t = W_{IH}x_t + W_{HH}h_{t-1} + b_h \tag{1.55}$$

$f_H(.)$ is the activation function for the hidden layer, and $b_h$ is the bias vector of the hidden units. These hidden units are then connected to the output layer through weighted connections $W_{HO}$.

To address the "long-term dependencies" issue in RNNs, the **Long Short-Term Memory (LSTM)** cell was introduced as an architectural extension of the standard recurrent units. This advancement enhanced the memory capacity of standard recurrent cells by incorporating gates into the architecture. LSTMs have since been modified and popularized by researchers. Variations of LSTMs include those without a forget gate, those with a forget gate, and those featuring a peephole connection [181–183]. Generally, when the term "LSTM cell" is used, it refers to the version with a forget gate (Fig. 1.41).

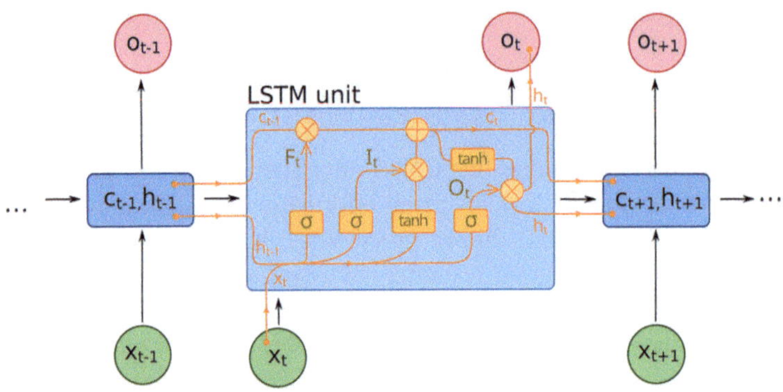


**Fig. 1.41** The LSTM architecture (https://commons.wikimedia.org/wiki/File:Long_Short-Term_Memory.svg)

The equation for the LSTM without a forget gate is expressed as follows:

$$\begin{aligned} i_t &= \sigma\left(W_{ih}h_{t-1} + W_{ix}x_t + b_i\right), \\ \zeta_t &= \tanh\left(W_{\zeta h}h_{t-1} + W_{\zeta}x_t + b_{\zeta}\right) \\ c_t &= c_{t-1} + i_t\zeta_t \\ o_t &= \sigma\left(W_{oh}h_{t-1} + W_{ox}x_t + b_o\right) \\ h_t &= o_t \tanh\left(c_t\right) \end{aligned} \tag{1.56}$$

To further enhance LSTMs, Gers and Schmidhuber introduced peephole connections, which improve the gates' access to essential information that may influence the network's performance [184].

The **transformer model** was introduced as a complete shift from the previous sequential processing methods utilized by RNNs and their variants. It leverages attention mechanisms to process input data in parallel, resulting in faster and more efficient computing. The architecture of a transformer model consists of a stack of identical modules, each containing two sub-layers: a multi-head self-attention mechanism and a position-wise fully connected feed-forward network [185, 186]. This architecture enables transformer models to capture dependencies between inputs and outputs, regardless of their position in the sequence.

**Transformer-based architectures** have shown significant promise in handling large-scale and complex datasets because they can effectively capture long-range dependencies, simultaneously consider multiple relevant data sets, and adaptively focus on important features. Additionally, transformer architectures have demonstrated great potential in natural language processing (NLP). NLP is a branch of machine learning focused on linguistics and the development of language models. Language Modeling (LM) determines the probability of an ordered sequence of words occurring in a sentence using probabilistic and statistical methods. Initially, language models were based on RNNs; however, due to issues such as vanishing and exploding gradients for long sequences, they were superseded by LSTMs and GRUs for improved performance. Despite advancements in LSTMs, they struggled with reliable comprehension of larger sequences. Consequently, the transformer architecture was developed to overcome the context limitations of LSTMs through its attention mechanism, providing greater throughput by processing inputs in parallel without sequential dependencies [187–190].

**Attention** is a complex cognitive function that plays a crucial role in human perception. Humans do not process all information simultaneously; instead, they concentrate on certain details when necessary, ignoring irrelevant information. **Attention mechanisms** can be categorized based on how they are generated. The first category is bottom-up unconscious attention, also known as saliency-based attention, which is driven by external stimuli. For example, people are likely to notice louder voices during conversations. This is analogous to **max-pooling** and **gating mechanisms** in deep learning, which prioritize more significant and relevant values for the next steps. The second category is top-down conscious attention, also known as focused attention. This type of attention serves a specific, predetermined purpose and is task-dependent [191]. Most deep learning attention mechanisms are designed to focus on specific tasks, thus fall under focused attention.

**Focused attention mechanisms** act as resource allocation methods to address information overload. They are particularly useful when computing power is limited, as they allow for the processing of more relevant and important information with fewer resources. Various applications of attention mechanisms include saliency-based visual models that extract local, low-level visual features to identify significant regions, as well as applications in image classification using neural networks and simultaneous translation and alignment tasks in machine translation. They are commonly used in neural architectures for tasks such as generating captions for images, text classification, translation, action recognition, image analysis, and speech recognition.

In addition to improving performance, attention mechanisms can also elucidate the behavior of complex neural architectures. In recent years, neural networks have made substantial progress in various sectors, including finance, materials, meteorology, medicine, autonomous driving, human-computer interaction, behavioral analysis, and emotional detection. However, the lack of interpretability poses challenges regarding ethical and practical issues [192]. While the interpretability of deep learning is still debated, attention mechanisms can provide relevant explanations to a certain extent.

**Graph Neural Networks (GNNs)** are a class of models designed to operate on data structured as graphs, which consist of nodes (representing entities) and edges (representing relationships). GNNs enable each node in a graph to iteratively gather information from its neighbors, allowing the model to learn not only from the node's own features but also from the structure and attributes of its local neighborhood. This message-passing framework makes GNNs particularly suitable for analyzing systems where relationships and connectivity matter more than simple linear ordering, such as in computer networks, social graphs, recommendation engines, and biological systems. At each layer, a node's representation is refined by aggregating information from its neighboring nodes, capturing both local and higher-order structural dependencies.

For example, in a communication network, a GNN can model the traffic load of a router by taking into account the activities of nearby routers. In a cybersecurity application, GNNs can identify suspicious behavior patterns by analyzing connections among devices over time. Moreover, several companies, such as Pinterest and Alibaba, utilize GNNs to enhance content and product recommendations through the analysis of user-item interaction graphs. GNNs also play a significant role in drug discovery, where molecules are represented as graphs with atoms and bonds [193]. They can predict the biological activity of compounds or discover new drugs by learning from molecular structures. In the financial sector, transactions between users and accounts form a graph, and GNNs can help identify suspicious clusters or patterns indicative of fraudulent behavior.

Traditional data mining and machine learning algorithms rely on statistical models trained on pre-collected labeled or unlabeled training data to predict future outcomes. **Semi-supervised classification** addresses the issue of insufficient labeled data to create effective classifiers by leveraging a large amount of unlabeled data alongside a limited quantity of labeled data. Variations of supervised and semi-supervised learning have been explored for defective datasets; for instance, Zhu and Wu have investigated solutions for noisy class label problems [194]. When more tests may be conducted on future samples, Yang et al. have taken cost-sensitive learning into account.

However, most of these approaches assume that labeled and unlabeled data share identical distributions. In contrast, transfer learning allows for the use of multiple domains, tasks, and distributions for testing and training. Numerous real-world examples of transfer learning exist. For instance, learning to identify apples can aid in distinguishing pears. The ability to effectively apply previously acquired knowledge to solve new problems more quickly or efficiently is the driving force behind research in transfer learning. The necessity for lifelong machine-learning techniques that retain and reuse previously learned knowledge was discussed during a NIPS-95 workshop on "Learning to Learn."

Today, **transfer learning** techniques are employed in various prestigious fields, including machine learning (ICML, NIPS, ECML, AAAI, and IJCAI), data mining (ACM KDD, IEEE ICDM, and PKDD), and applications of both machine learning and data mining (ACM SIGIR, WWW, and ACL). Pretrained models serve as a foundation for feature extraction or are fine-tuned with task-specific data. This approach significantly reduces training time and resource consumption while often enhancing performance.

Transfer learning has a wide range of applications in sequential machine learning as well. For example, a graph-based method for detecting previously encountered games has been developed to speed up reinforcement learning on variations of games already played. This methodology has been utilized to automate domain mapping for value function transfer. A novel approach for transferring knowledge between completely separate feature spaces has been proposed in translated learning, which involves learning a mapping function that bridges features from two different domains [195]. Additionally, Li et al. have applied transfer learning to address issues of sparsity and cold starts in collaborative filtering, discovering a Rating-Matrix Generative Model (RMGM) as a shared rating-pattern mixture model based on latent user and item-cluster variables.

There are numerous industries employing these techniques. In the medical field, CNNs pretrained on ImageNet can be fine-tuned for classifying medical scans, such as chest X-rays or MRIs, offering high accuracy even with small datasets. There is also a wide application for transfer learning in the self-driving vehicle industry, where perception modules in autonomous systems utilize pretrained models to enhance their performance.

## 1.9 Fuzzy Systems and Their Applications

**Fuzzy logic** is a type of many-valued logic that focuses on reasoning that is approximate rather than fixed and exact. Unlike classical binary logic, where variables must be either true or false, fuzzy logic allows variables to have truth values ranging from 0 to 1. This flexibility enables the modeling of concepts that are inherently vague or subjective, such as "tallness" or "warmth."

The concept of fuzzy logic was first proposed by Lotfi A. Zadeh in 1965 as an extension of classical logic designed to accommodate the idea of partial truth. He introduced the notion of fuzzy sets, in which elements can have degrees of membership that range between 0 and 1, rather than being solely included or excluded as in classical sets. At the core of fuzzy logic is the membership function. This function quantifies how much a particular element belongs to a fuzzy set. For instance, when defining the fuzzy set "warm temperatures," a temperature of 22 °C might have a membership value of 0.7, indicating that it is considered moderately warm [196]. Membership functions can take various shapes, like triangular, trapezoidal, or Gaussian, depending on the specific application and the desired sensitivity.

Fuzzy logic systems use a set of rules known as **fuzzy rules** to make decisions based on the degrees of truth associated with input variables. These rules are typically formulated as "IF-THEN" statements. For example, "IF the temperature is high, THEN the fan speed is fast." The system evaluates all relevant rules and combines the results to produce a final output, which is then defuzzified to yield a specific action or decision.

In real life, we often make decisions based on incomplete, imprecise, or subjective information. For example, we might describe the weather as "rather warm" or a person's mood as "slightly off." Such descriptions capture nuances that traditional binary logic, which categorizes statements as purely true or false, fails to address. Classical logic adheres to the principle of bivalence, which asserts that a proposition is either true or false. This principle works well for problems that are clearly defined, but it struggles with the vagueness and uncertainty characteristic of most real-world situations [197]. For instance, the question of whether someone is "tall" cannot be answered simply with "true" or "false," since perceptions of height can vary significantly among individuals and cultures. Classical logic cannot accommodate such variations in truth.

To overcome these limitations, alternative logical systems have been developed. Fuzzy logic, by allowing for degrees of truth, enables more nuanced reasoning that aligns closely with human thought processes. By accommodating partial truths and effectively handling uncertainty, fuzzy logic offers a more robust framework for modeling complex systems and making decisions in ambiguous contexts.

### *1.9.1 Fundamentals of Fuzzy Logic*

In traditional logic systems, often referred to as **crisp logic**, statements are evaluated in a binary manner: they are either completely true or completely false. This approach works well for straightforward scenarios but falls short in handling the ambiguity present in many real-world situations.

Fuzzy logic addresses this limitation by allowing for degrees of truth. Instead of assigning a strict true or false value, fuzzy logic permits values between 0 and 1, representing a continuum of truth. For example, while crisp logic would classify a person as either "tall" or "not tall," fuzzy logic allows for a person to be "somewhat tall," with a membership value of, say, 0.7 in the "tall" category. This nuanced approach enables systems to manage uncertainty and partial truths, making fuzzy logic particularly useful in fields such as control systems, decision-making, and artificial intelligence.

In fuzzy logic, linguistic variables are those whose values are represented by words or phrases in natural language rather than numerical values. These variables facilitate a more intuitive representation of information, closely mirroring human reasoning. For instance, consider the variable "temperature." Instead of assigning a specific numerical value, we might describe temperature using terms like "cold," "warm," or "hot." Each of these terms corresponds to a fuzzy set with an associated membership function, which defines the degree to which a particular temperature value belongs to that category [209]. This approach simplifies the process of formulating rules and making decisions based on imprecise or subjective information, which is common in real-world scenarios.

Fuzzy logic extends classical logical operations to accommodate degrees of truth. The fundamental operations—AND, OR, and NOT—are redefined to handle fuzzy values:

- **Fuzzy AND (Conjunction):** This operation determines the degree to which both conditions are satisfied. Typically, the minimum value of the involved membership degrees represents this. For example, if condition A has a truth value of 0.7 and condition B has a truth value of 0.5, then A AND B would have a truth value of min (0.7, 0.5) = 0.5.
- **Fuzzy OR (Disjunction):** This operation assesses the degree to which at least one condition is satisfied. Typically, the maximum value of the involved membership degrees represents this. Using the previous example, A OR B would have a truth value of max (0.7, 0.5) = 0.7.
- **Fuzzy NOT (Negation):** This operation calculates the degree to which a condition is not satisfied. It is typically represented by subtracting the membership degree from 1. If condition A has a truth value of 0.7, then NOT A would have a truth value of 1–0.7 = 0.3 (Fig. 1.42).

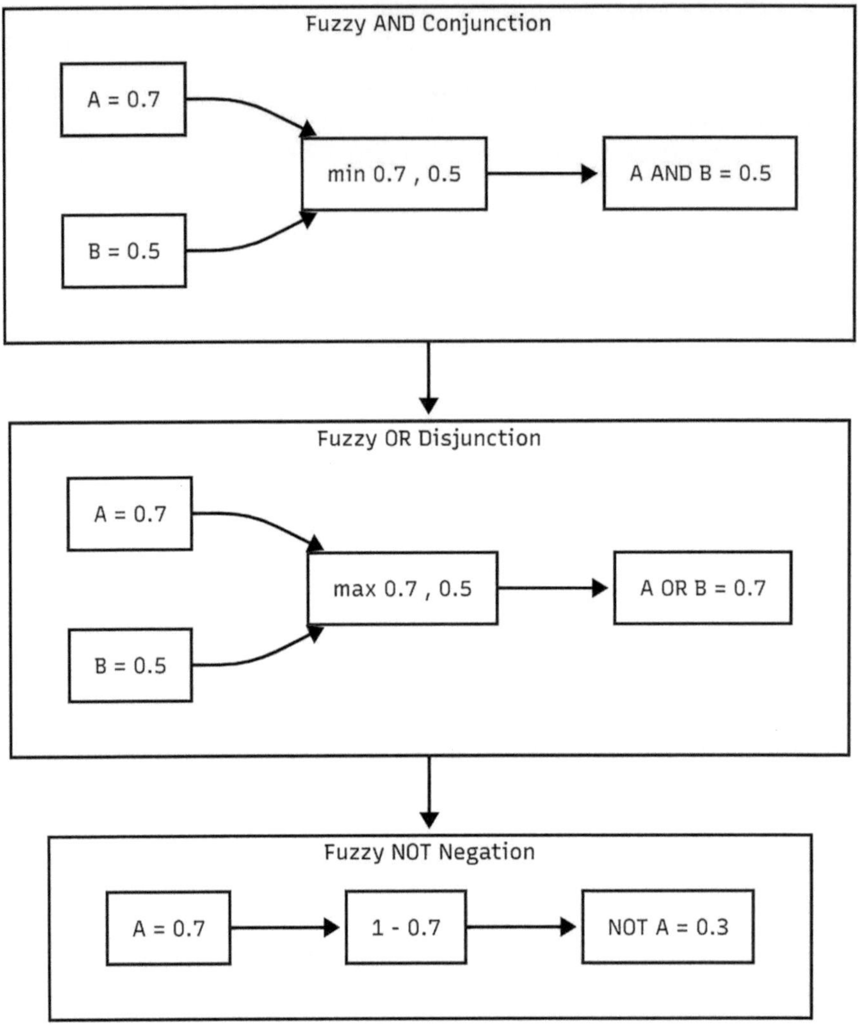


**Fig. 1.42** Fuzzy Logic Operations

In classical set theory, an element either belongs to a set or it does not—membership is binary. However, many real-world concepts are not so clear-cut. For example, consider the concept of "tallness." At what exact height does someone become "tall"?

A fuzzy set allows elements to have degrees of membership that range from 0 to 1. This means an individual can partially belong to a fuzzy set to a certain extent. Formally, a fuzzy set A in a universe of discourse X is defined by a membership function $\mu A(x)$: $X \rightarrow [0,1]$, where each element x in X is assigned a membership value $\mu A(x)$ that indicates its degree of membership in set A.

The **membership function** (MF) is a fundamental component of fuzzy logic, defining how each point in the input space is mapped to a membership value between

0 and 1. The shape of the MF determines how each input value is fuzzified. Common types of membership functions include [198]:

The **triangular MF** is defined by three parameters: a, b, and c, where a < b < c. The function increases linearly from a to b and decreases linearly from b to c. This membership function is simple and computationally efficient, making it popular in real-time systems.

The **trapezoidal MF** is an extension of the triangular MF, defined by four parameters: a, b, c, and d, where a < b ≤ c < d. It has a flat top, which represents full membership over a range of values. This membership function is applicable when a range of values equally represents a particular concept.

The **Gaussian MF** is characterized by its smooth, bell-shaped curve and is defined by two parameters: c (the center) and σ (the width). This MF is differentiable and provides smooth transitions, making it suitable for control systems that require gradual changes (Fig. 1.43).

## *1.9.2 Fuzzy Inference Systems*

Fuzzy Inference Systems (FIS) offer an alternative to traditional rule-based systems by utilizing fuzzy sets. They can work with imprecise data to emulate human thinking. FIS operates with fuzzy data to represent human reasoning, making it broadly applicable across various industries. Their flexibility and efficiency have been demonstrated in numerous fields, enabling human intuition and computation to transform decision-making in response to modern demands.

FIS are among the most recognized applications of fuzzy logic and fuzzy set theory. They serve as the foundation for fuzzy logic systems, which take input values and produce output values based on a set of rules. FIS are widely applied in control systems, data classification, and decision-making systems, especially in scenarios involving uncertainty or imprecise information. Unlike classical logic systems that operate on clear binary true or false values [199], FIS mirrors human intuition by

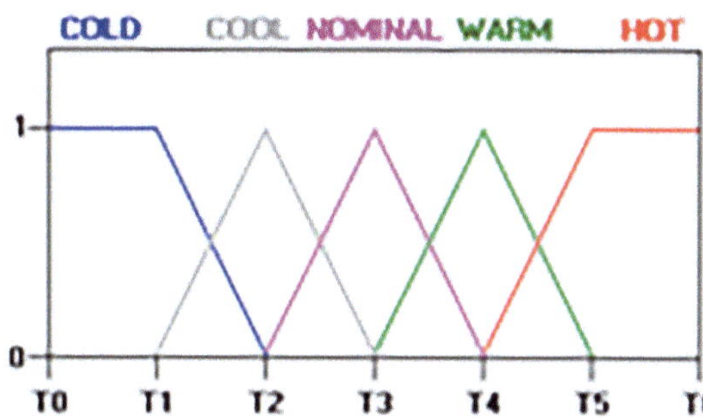


**Fig. 1.43** Membership Function Representation (https://commons.wikimedia.org/wiki/File:Fuzzy_control_-_definition_of_input_temperature_states_using_membership_functions.png)

allowing for degrees of truth and falsehood, with values ranging from 0 to 1. The core concept of FIS involves accepting fuzzy input values, processing them through a series of rules and an inference engine, and generating fuzzy output values.

**Structure of a Fuzzy Inference System**

A Fuzzy Inference System consists of several components that facilitate the transformation of clear inputs into clear outputs through input fuzzification, a rule base, an inference mechanism, and defuzzification.

FIS is composed of four key components:

1. Fuzzification: Converts clear inputs into fuzzy values.
2. Rule Base: A set of if-then rules that illustrate how the inputs relate to the output.
3. Inference Engine: Processes the fuzzy input and applies the rules from the rule base.
4. Defuzzification: Converts the fuzzy output back into a clear value, resulting in the final output (Fig. 1.44).

**Fuzzification** is the initial step in a fuzzy inference system. It can be thought of as taking precise, textbook-perfect numbers and blending them together to produce fuzzy sets. Membership functions act like recipes, determining how "fuzzy" each input becomes on a scale from 0 to 1. The choice of membership function shape—be it triangular, trapezoidal, Gaussian, or another type—significantly impacts the overall system's performance [200]. The **rule base** contains fuzzy if-then rules, which are formulated either through expert knowledge or data-driven methods. For example, one rule might state that if the temperature is high and the humidity is low, then the fan speed should be increased. These rules govern the logic of the fuzzy inference engine. The **inference engine** applies the fuzzy rules to the fuzzified inputs, generating fuzzy outputs. It employs logical operations—such as AND, OR, and NOT—to interpret the fuzzy values of the input data concerning each rule. This may involve different approaches, such as the **Mamdani or Sugeno methods**, to effectively aggregate rule outputs. **Defuzzification** is the final stage in a fuzzy inference system. It converts the fuzzy output sets generated by the inference engine into a single crisp output. Common defuzzification methods include the centroid method, bisector method, and the maximum membership principle. The selection of a defuzzification method can significantly affect the responsiveness and stability of the control system.

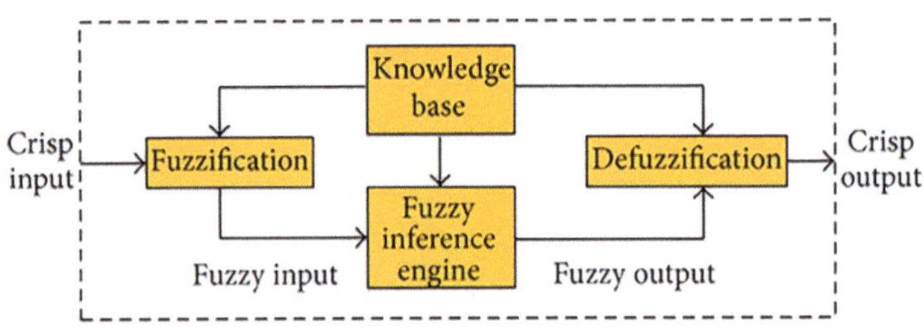


**Fig. 1.44** Fuzzy Inference System (https://commons.wikimedia.org/wiki/File:Fuzzy-logic-controller-block.jpg)

The **Mamdani inference system**, developed by Ebrahim Mamdani in 1975, the Mamdani approach is a widely used method in control systems. It aligns closely with natural human reasoning by considering fuzzy sets for both inputs and actions. Mamdani rules typically have the form: "IF x is A AND y is B THEN z is C," where A, B, and C are fuzzy sets. The outputs are aggregated and then defuzzified to yield a crisp result.

One advantage of Mamdani systems is their interpretable nature; rules using linguistic variables make it easier to follow and adjust the system. However, a disadvantage is the complexity and computational cost of the defuzzification step when dealing with a large number of rules.

**Sugeno inference systems**, developed by Takagi, Sugeno, and Kang in the 1980s, differ from Mamdani systems in that they use linear functions or constants as rule consequents instead of fuzzy sets. A typical Sugeno rule looks like this: "IF x is A THEN z is k," where k is a constant (value). This key difference makes Sugeno systems more suitable for certain applications, particularly in optimization processes and when a weighted average of output is needed.

### *1.9.3 Fuzzy Control Systems*

**Fuzzy control systems** utilize the principles of fuzzy logic along with automatic control theory to manage processes that are difficult to model mathematically. These controllers interpret each input value with fuzzy logic and adjust the control actions accordingly, which provides them with significant flexibility and adaptability in uncertain environments [201].

Fuzzy logic extends classical logic by permitting partial truths, allowing systems to function effectively even with incomplete or uncertain information. Fuzzy control systems can make continuous adjustments in their responses while minimizing oscillations in output.

Traditional **Proportional-Integral-Derivative (PID) controllers** require precise mathematical models and perform optimally under linear, time-invariant conditions. However, PID controllers struggle with nonlinear systems and uncertainties. In contrast, fuzzy controllers employ rules that recognize system complexities and better handle uncertainties. Since it is often impossible to obtain an exact mathematical model, fuzzy controllers offer smoother and more graded responses than PID controllers. This is particularly evident in scenarios such as controlling a robotic arm under unexpected resistance or reacting to changes in load.

**Designing a Fuzzy Controller**

Designing a fuzzy controller typically involves five steps:

1. Identify input and output variables.
2. Define the membership functions.
3. Create a rule base.
4. Implement the inference engine and defuzzification.
5. Test and fine-tune the system.

For example, in a temperature control application, the input variables might include temperature error and the rate of change of that error, while the output variable would represent the control signal to the heater. The membership functions would convey linguistic terms (such as "cold," "warm," or "hot"), and the rules (for example, "IF the temperature is cold AND falling rapidly, THEN the heater setting is high") would dictate the control action.

**Fuzzy control systems** have numerous practical applications, particularly in domains where traditional modeling techniques fall short. Major application areas include robotics, HVAC (Heating, Ventilation, and Air Conditioning) systems, and automotive systems. In robotics, fuzzy logic facilitates the management of robot motion and behavior by interpreting sensor data to inform decisions about movement. For vehicles equipped with fuzzy logic, a fuzzy controller effectively manages both obstacle avoidance and path planning, enabling robot behavior that mimics human responses. In HVAC applications, fuzzy controllers can comprehend the implications of temperature errors, as well as variables such as occupancy and humidity, enabling better management of temperature and airflow to achieve desired outcomes. Unlike traditional HVAC systems that merely turn on and off, fuzzy HVAC controllers can implement gradual and complex adjustments, leading to increased comfort and energy savings [202].

Within automotive systems, fuzzy logic is applied to various functionalities, such as automatic transmission, braking, and cruise control systems. For instance, fuzzy systems can optimize vehicle gear shifts based on road traffic and conditions, as well as driver behavior, resulting in smoother rides and improved fuel efficiency (Fig. 1.45).

Fuzzy inference systems can enhance **medical diagnosis** by effectively processing uncertain patient information and imprecise symptoms. Research has demonstrated that fuzzy expert systems can predict whether a COVID-19 patient will require admission to an ICU based on various clinical indicators. By employing fuzzy diagnostic systems, we can enhance diagnostic accuracy, particularly in situations where data are uncertain or when patients do not provide complete information about their symptoms.

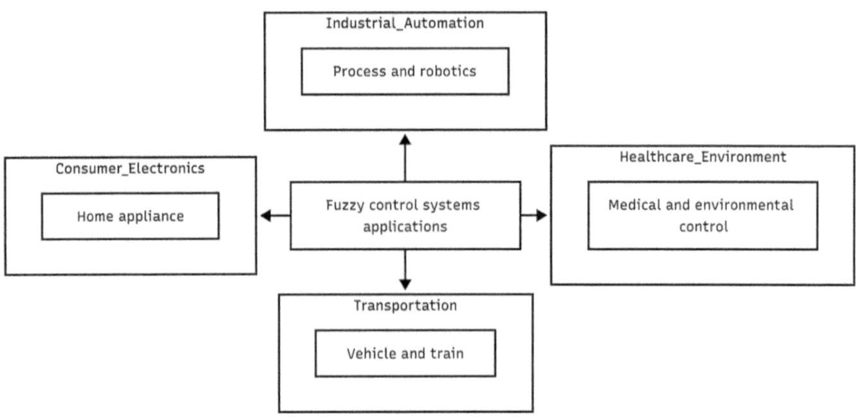


**Fig. 1.45** Applications of Fuzzy Control Systems

Fuzzy systems play a crucial role in managing urban traffic flow by monitoring vehicle counts, levels of congestion, and traffic signal timing. They can dynamically optimize traffic light timing, thereby minimizing delays and enhancing overall traffic flow. Weather prediction is inherently uncertain, making it a suitable application for fuzzy logic. Fuzzy systems can utilize imprecise data related to factors such as humidity, temperature, and wind velocity, while also incorporating pressure and other elements. This approach reflects real-world conditions more accurately than traditional rigid models. In the financial services sector, fuzzy systems are valuable tools for assessing credit risk. These innovative systems can effectively represent fuzzy data, capturing the complex communication between various financial variables. Furthermore, they offer a more sophisticated scoring system than binary models, leading to fairer and safer lending practices.

**Hybrid intelligent systems** combine fuzzy logic with neural networks. This approach allows the system to learn from data input while incorporating fuzzy variables within the context of that data. As a result, these systems can adapt more effectively and perform better in complex situations. Fuzzy systems require well-defined rules and membership functions; if poorly designed, they can generate inaccurate outputs. Furthermore, fuzzy systems do not possess the ability to learn or adapt based on knowledge gained from previous decisions. Unlike machine learning models, fuzzy systems do not have the capability to modify their decisions over time. Future trends will highlight intelligent control within innovative systems that integrate AI and the Internet of Things (IoT). This combination aims to minimize human error and create a more intelligent, coordinated, and responsive control system. Applications such as smart cities, autonomous vehicles, and energy management are expected to see significant growth in demand.

## 1.10 Modern AI and Emerging Technologies

AI has evolved significantly from its origins in science fiction. It has become a fundamental element driving technological advancement across the globe. The concept of AI emerged in the mid-twentieth century, leading to the development of machines capable of mimicking human traits, including language processing, learning, and problem-solving. Although the initial AI systems were rudimentary, they incorporated elements of digital cognition and automation, paving the way for future systems with far greater capabilities.

In the twenty-first century, we are experiencing an unprecedented acceleration of technological progress, with AI serving as a pivotal force in innovation. Society is undergoing significant changes, as industries and economic models become increasingly interconnected. AI is enhancing the controllability and accessibility of devices, thanks to improved data availability and computational power [203]. The transformations brought about by AI extend beyond mere technological advancements; the influence of AI is reshaping relationships among humans, machines, and important global issues.

### 1.10.1 Generative AI

**Generative AI** refers to algorithms that create new content based on patterns learned from a dataset. Unlike traditional machine learning models that focus on classification or prediction, generative models produce content such as images, audio, or text. The use of Generative AI has grown significantly in recent years, driven by advancements in computational capabilities and an increasing demand for realistic content generation. Key technologies in this area include **Generative Adversarial Networks (GANs)** and, more recently, **Diffusion Models**.

GANs consist of two neural networks: the generator and the discriminator. These networks are organized to create a constant zero-sum game. The generator creates data that mimics the training set, while the discriminator evaluates whether the examined data is real or fake. Through repeated training cycles, both networks continuously improve. As the discriminator provides feedback, the generator learns to produce data that can deceive the discriminator. This adversarial training enables the model to learn complex data distributions effectively. Recently, diffusion models have gained attention in applications such as medical imaging predictions. These models serve as efficient methods for image generation, offering high precision and flexibility. Their adaptability helps researchers in constrained environments, establishing trust in the available models.

Beyond medical imaging, diffusion models have demonstrated exceptional performance in high-resolution image synthesis, inpainting, and text-to-image generation. When applied to intuitive tasks, diffusion models generate outputs that exhibit a higher degree of realism. However, the challenge remains in finding plausible outputs that embody a wide range of characteristics while incorporating fine-grained control over realism. Compared to GANs, Diffusion Models generally provide more stable training and higher-fidelity outputs, although at a greater cost in terms of time and computation [204]. Recent advancements, including Denoising Diffusion Probabilistic Models (DDPM) and latent diffusion architectures, have addressed various issues and improved performance trade-offs, making this technology more accessible for practical applications across diverse industries, including entertainment and scientific visualization.

**Diffusion models**, particularly DDPMs, operate by converting noise into organized data through a **Markov Chain**. Initially overlooked, these models have gained popularity as text-to-image generators, thanks to advancements in technologies like DALL·E 2 and Stable Diffusion.

While GANs produce detailed images with layers that are undetectable by the human eye, the added complexity in Spectrum Models enhances image stability and increases the difficulty of samples while reducing the number of training parameters. Diffusion Models, despite requiring a longer training time, improve the stability of the generation process.

Although susceptible to mode collapse, Spectrum Models need more time for their iterative denoising protocols, which enhance both the stability and variability of the generated images. Despite their potential, these models require further evaluation before a wide public release. Excessive precision raises concerns about content authenticity and the potential for misinformation, leading to deeper discussions around ethical frameworks related to sensitive areas, such as personal privacy and public security [205]. Using GANs in cost-sensitive machine learning tasks, along with generating images from text, can benefit low-resource environments. This approach can enhance medical imaging, while DDPMs, combined with specific patterns of noise, help transform input images and enable the creation of highly detailed content using easily generated components (Table 1.14).

### *1.10.2 Federated Learning*

**Federated learning** is a transformative approach to machine learning. Instead of sending all the data to a central server, which poses significant privacy risks, each device processes the data locally. Then, the devices send their model updates back to a central server. Google pioneered this concept in 2016, seemingly anticipating the impending privacy issues in sectors like finance and healthcare.

In this architecture, although there is a central server, the majority of processing occurs on edge devices, such as mobile phones or hospital servers. These devices train their models locally and then contribute their updates to a collective pool. The technique known as FedAvg plays a crucial role by combining these updates efficiently, allowing the system to scale without compromising performance. When integrated with neural networks, federated learning can yield accurate predictions while maintaining user privacy.

However, concerns exist regarding privacy and security. Sophisticated hackers can exploit vulnerabilities through model inversion or poisoning attacks [206]. As a result, techniques like secure aggregation, homomorphic encryption, and differential privacy are critical to enhancing the security of federated learning systems. The applications for federated learning are vast. For instance, doctors in different hospitals can collaborate on diagnostics without requiring questionable data-sharing agreements, and banks can detect fraud while remaining compliant with regulatory requirements. As the technology continues to evolve, it is likely to become the backbone of any ethical, privacy-first AI system. The future looks promising.

Federated learning can vary based on the structure of datasets or the characteristics of client nodes. It is typically classified into several categories:

- Cross-device: This involves devices with unstable connectivity and limited computing resources, such as mobile phones and Internet of Things (IoT) devices.
- Cross-silo: This involves a small number of servers or data centers with stable connectivity and sufficient computational power to handle large volumes of data.
- Horizontal: In this setup, client node datasets share the same features or structure but consist of different samples.
- Vertical: This involves client node datasets that share the same samples but differ in structure or features (Fig. 1.46).

Table 1.14 Comparison of GANs and Diffusion Models

| Aspect | Generative Adversarial Networks (GANs) | Diffusion Models |
|---|---|---|
| Core principle | Learn to generate data by playing a minimax game between a generator and a discriminator | Learn a forward diffusion process that gradually adds noise and a reverse process that denoises to recover data |
| Training objective | Adversarial loss (e.g., Jensen–Shannon, Wasserstein) where the generator tries to fool the discriminator | Minimize the difference between predicted and true noise (score-matching or denoising loss) across many diffusion steps |
| Sampling process | A single forward pass through the generator produces a sample | Iterative reverse diffusion process (hundreds to thousands of denoising steps) |
| Training stability | Can be unstable, prone to mode collapse, and sensitive to hyperparameters | Generally, more stable and less prone to mode collapse |
| Sample quality | Very sharp, high-fidelity images when training succeeds | High-quality and diverse samples; fidelity improves with more denoising steps |
| Speed of generation | Fast once trained (one pass) | Slow due to many reverse steps (though accelerated samplers exist) |
| Data efficiency | Requires large datasets but can be adapted to smaller sets with careful regularization | Typically needs large datasets and a long training time |
| Strengths | High-resolution image synthesis, real-time generation, many architectural variants (StyleGAN, CycleGAN) | Excellent diversity, strong likelihood estimation, state-of-the-art image and audio generation |
| Limitations | Mode collapse, training instability, difficulty estimating likelihood | Computationally expensive sampling, slower inference |
| Typical applications | Image-to-image translation, super-resolution, video synthesis, data augmentation | Text-to-image models (e.g., stable diffusion), speech synthesis, and high-fidelity image generation |

Federated learning reduces the need to access or transfer massive datasets. Since data is stored and processed locally, this approach helps organizations comply with data protection regulations. While implementing federated learning for real-world applications can be complex, various frameworks are available to train models on decentralized data and streamline workflows between servers and clients.

### *1.10.3 Explainable AI*

Ridley (2022) defines "**explainable artificial intelligence (XAI)**" as a subfield of AI that offers explanations for the predictions, recommendations, and decisions made by intelligent systems. The importance of understanding AI systems, along with their transparency and interpretability, has never been greater, as AI technologies become increasingly sophisticated and integrated into decision-making across

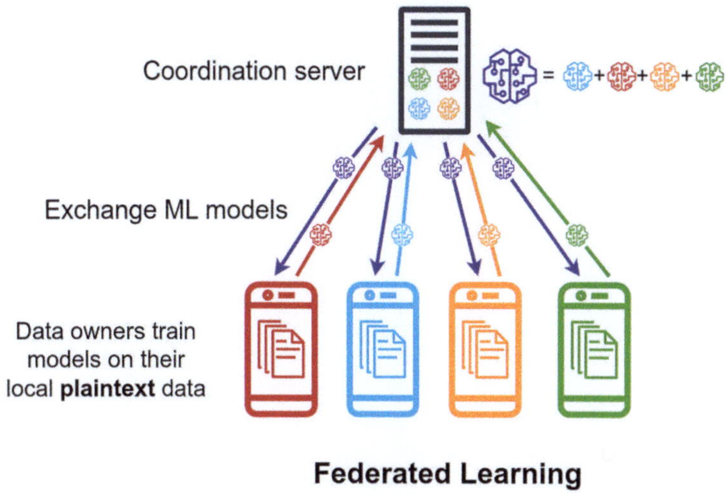


**Fig. 1.46** Federated Learning (https://commons.wikimedia.org/wiki/File:Federated_learning_protocol.png)

various sectors, including medical diagnosis and legal risk analysis. XAI encompasses a range of strategies designed to clarify the behaviors and decisions of AI systems to human users with diverse backgrounds. Unlike traditional black-box decision systems, XAI seeks to explain not just what a system predicts, but, more importantly, why it makes those predictions.

XAI methods can generally be categorized into two groups: **ante-hoc** and **post-hoc explainability.** Ante-hoc models, such as decision trees and rules-based models, are naturally interpretable. In contrast, post-hoc methods, including SHAP (SHapley Additive exPlanations), LIME (Local Interpretable Model-agnostic Explanations), and Grad-CAM (Gradient-weighted Class Activation Mapping), provide interpretability after the model has been trained [207]. These tools provide insights into feature importance, decision boundaries, and internal activations, making the black box transparent without compromising model complexity.

A novel approach involves **integrating XAI with generative models**, such as GANs and diffusion models. For example, Explainable GANs (XGANs) attempt to map latent representations to interpretable features, allowing users to visualize how inputs translate into synthetic outputs. In the context of diffusion-based structures, interpretability can be achieved by examining the denoising process and monitoring certain latent paths during sampling. These developments enhance user trust and encourage responsible use of generative tools.

In clinical settings, XAI has been utilized to explain AI-enabled radiology diagnostics. Techniques such as heatmaps and saliency maps illustrate which features are relevant in medical images. In the finance sector, XAI is applied to credit scoring models to highlight which attributes and their significance influence loan approval decisions. Companies like Google and IBM have also developed XAI-driven technologies within their AI platforms, enabling stakeholders and regulators to more effectively review and audit decision-making processes (Table 1.15).

**Table 1.15** Overview of XAI Techniques

| Technique | Scope | Key Idea | Typical Models | Output Form | Strengths | Limitations |
|---|---|---|---|---|---|---|
| LIME (local interpretable model-agnostic explanations) | Local | Approximates a model's prediction around a single instance using a simple surrogate model (e.g., linear regression) | Any black-box model | Feature importance for one prediction | Model-agnostic, intuitive | Requires many perturbations; explanations can vary with sampling |
| SHAP (SHapley additive exPlanations) | Local + global | Uses Shapley values from cooperative game theory to attribute contributions of each feature | Any black-box model | Consistent feature importance values | Theoretically sound, additive, can aggregate for a global view | Computationally expensive on high-dimensional data |
| Partial dependence plots (PDP) | Global | Shows the average effect of one or two features on the predicted outcome while marginalizing others | Tree ensembles, linear models, neural networks | 2D or 3D plot of feature vs. prediction | Simple global insight | Assumes feature independence; can be misleading with strong interactions |
| Individual conditional expectation (ICE) | Local + global | Shows how prediction changes across a feature for each instance | Same as PDP | Multiple instance-level curves | Reveals heterogeneity hidden in PDP | Hard to read with many instances |
| Feature importance (Gini, gain) | Global | Measures the contribution of each feature to model accuracy or loss | Tree-based models (RF, XGBoost, LightGBM) | Ranked list of features | Fast, built-in for trees | Only global; biased toward high-cardinality features |

| | | | | | | |
|---|---|---|---|---|---|---|
| Counterfactual explanations | Local | Identify minimal changes to input features that flip a model's decision | Any model with differentiable or query access | Example of modified input | Actionable guidance | May produce unrealistic or infeasible counterfactuals |
| Saliency/gradient maps | Local | Use input gradients to show which pixels or tokens most affect the prediction | CNNs, transformers | Heatmap over input | Visual, model-specific | Sensitive to noise; gradients can saturate |
| Integrated gradients | Local | Accumulate gradients along a path from baseline to input | Deep neural networks | Feature attribution scores | Addresses gradient saturation | Requires a suitable baseline choice |
| Surrogate decision trees/rule lists | Global | Fit a simple interpretable model to approximate a complex one | Any black-box model | Decision rules or tree structure | Human-readable global rules | Approximation error may be high if a complex model is highly non-linear |

Numerous international and national regulatory frameworks highlight the need for explainable AI, including:

- The GDPR introduces a "right to explanation," requiring companies to provide users with meaningful information about automated decisions.
- The EU AI Act categorizes AI systems as unacceptable, high-risk, or low-risk, requiring interpretability for high-risk applications.
- The AI Bill of Rights in the U.S. emphasizes "notice and explanation" for automated decisions.
- The OECD Principles on AI and the European Commission's Ethical Guidelines advocate for AI systems that are transparent, fair, and accountable.

These regulations are more than just legal requirements; they act as guidelines for the ethical deployment of AI. They compel organizations to rethink their AI development processes, necessitating the documentation of models, model audits, and the integration of interpretability throughout the model's lifecycle. Moving forward, intelligent machines will likely be tasked with making decisions that impact human safety, psychology, and society. The pressing question is how to integrate societal, legal, and moral values into the technological development of AI. Some theories suggest enabling robots to learn ethics independently from humans, possibly through machine learning. Regardless of the approach taken, it is crucial to acknowledge that the future of humanity will inevitably include AI, making this one of the most pressing issues to tackle in the years to come.

### *1.10.4 Neuromorphic Computing*

**Neuromorphic Computing,** also known as neuromorphic engineering, refers to a computing approach that mimics the functioning of the human brain. This paradigm closely emulates the architecture and computational methods found in biological neural networks. Inspired by neurobiology, neuromorphic computing aims to replicate functionalities such as synaptic plasticity, event-driven processing, and sparse coding using silicon-based neurons and synapses. Although the concept was developed in the late 1980s, it has gained renewed interest as traditional von Neumann architectures have struggled with real-time, energy-efficient processing capabilities essential for AI systems [208]. The neuromorphic model is grounded in spike-timing dependent plasticity (STDP), a learning rule based on the synaptic changes observed in biological neural networks. This process is analogous to biological and genetic neural development and aligns with Hebbian learning, where an increase in synaptic strength results from one neuron persistently stimulating another (Fig. 1.47).

**Neuromorphic systems** are particularly effective in edge AI applications that require real-time processing at low power, such as robotics, auditory processing, and sensory-motor control. Examples of neuromorphic processors include Intel's Loihi chip and IBM's TrueNorth, both of which have demonstrated capabilities in

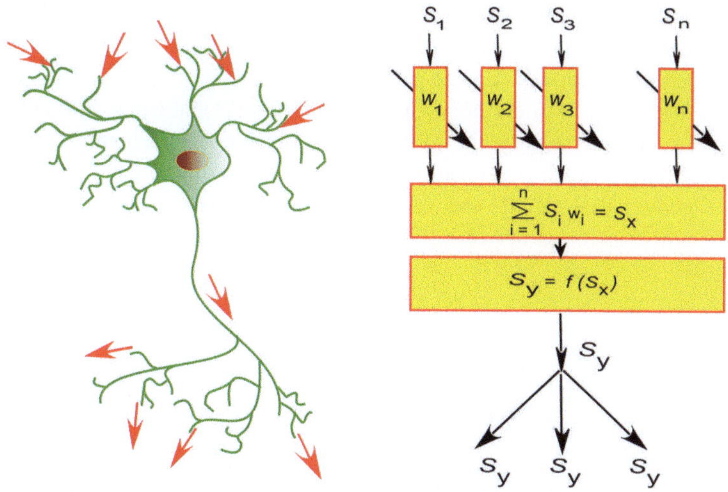


**Fig. 1.47** Comparison between biological neuron and artificial neuron (https://commons.wikimedia.org/wiki/File:ANN_neuron.svg)

pattern recognition, environmental interaction, and adaptive behavior. While computational costs may be associated with backpropagation or high-dimensional gradient descent, neuromorphic processing enables near real-time adaptation, distinguishing it from traditional computing methods.

In the field of biogenetics, neuromorphic computing holds significant relevance, as it mimics biological processes akin to those found in nervous systems, which are responsible for learning and memory. **Spiking neural network architectures** reflect the firing behaviors of biological neurons and the regulatory systems ingrained in biological genetics that influence neuroplasticity. For example, artificial synapses within neuromorphic chips can simulate **long-term potentiation** (LTP) and **long-term depression** (LTD), two genetic processes vital to memory and learning in living organisms. Furthermore, neuromorphic platforms provide computational tools that enable the simulation of biological neural networks involved in genetic diseases, such as epilepsy, autism, and Parkinson's. Researchers utilize these neuromorphic simulators to investigate how genetic mutations impact synaptic dynamics and the overall brain circuitry. In this way, neuromorphic technology not only replicates biological processes but also contributes to the understanding of biological systems.

One of the most intriguing aspects of neuromorphic computing is the potential to **bridge artificial intelligence and biological systems at a molecular level.** While neuromorphic systems are primarily designed to mimic biological neurons, they also strive to replicate core dynamics such as ion channel activity, neurotransmitter vesicle release, and synaptic density regulation—all of which are intricately regulated by complex gene networks in living organisms. This integration presents new opportunities for computational neuroscience, particularly in studies aimed at mimicking or simulating genetic disorders that affect brain function. Recent research has focused on the application of neuromorphic platforms to model gene expression pathways that influence neural plasticity. For instance, studies are emerging that suggest artificial synapses can be programmed to reflect epigenetic regulations associated with long-term memory. In this context, neuronal transcription linked to gene expression can modify the circuitry responsible for enduring structural or functional changes. Such possibilities enable robust modifications to the activity of artificial "genes" that imitate the roles of natural genes in regulating adaptation across the entire artificial network [209]. This includes the effects of brain-derived neurotrophic factor (BDNF) on synaptic development, neurogenesis, and neuroplasticity, among other relevant genes (Fig. 1.48).

### *1.10.5 Quantum Machine Learning*

**Quantum Machine Learning (QML)** exists at the intersection of two revolutionary fields: **quantum computing and artificial intelligence**. As classical machine learning approaches the physical and computational limits of silicon-based processors, QML presents itself as a promising framework capable of exponentially accelerating problem-solving through the principles of quantum mechanics. Unlike classical algorithms, quantum algorithms utilize phenomena such as superposition, entanglement, and quantum interference to process vast solution spaces in parallel, potentially redefining the boundaries of what is computationally feasible.

At its core, QML adapts traditional learning models, such as neural networks and support vector machines, into quantum frameworks. This often involves encoding data into quantum states via quantum feature maps and processing them using unitary transformations defined by quantum circuits. These methods rely heavily on linear algebra and tensor mathematics, which are also fundamental to classical neural networks. The similarities to neuromorphic systems are striking; just as spiking

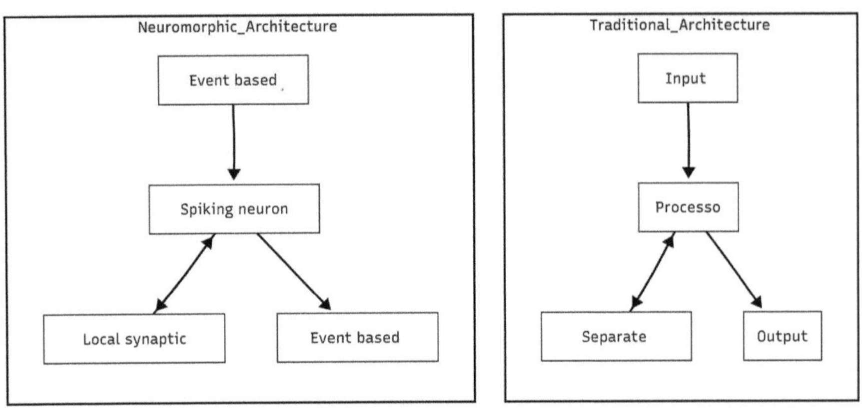


**Fig. 1.48** Traditional vs. Neuromorphic Architectures

neural networks aim to mimic the timing of biological neurons, quantum neural networks introduce probabilistic reasoning and state collapses that resemble the non-deterministic responses found in biological systems [210].

Though still in its early stages, QML has already demonstrated its value in complex applications such as drug discovery, financial modeling, and optimization problems. For example, in healthcare analytics, quantum-enhanced clustering and classification models have shown promise in analyzing genomic datasets with high-dimensional features—areas where traditional algorithms often struggle due to the curse of dimensionality. The efficiency of QML in exploring combinatorial states parallels the biological processes of genetic permutation and variation, emphasizing its relevance in genetics and bioinformatics.

Moreover, **federated learning and QML are not mutually exclusive**. Ongoing research is exploring how quantum communication protocols could facilitate decentralized quantum models to collaborate without compromising data privacy, thereby combining the benefits of federated learning architectures with the power of quantum acceleration. These concepts align with emerging advances in neuromorphic and generative architectures that also aim to maximize computational efficiency while adhering to ethical principles (Fig. 1.49).

The evolution of AI is increasingly characterized not by isolated inventions, but by the merging of technologies at conceptual, computational, and biological levels. The introduction of hybrid models, where multiple paradigms of AI converge within shared architectures, represents a significant shift from siloed systems to deeply interdependent, flexible intelligence systems. Hybrid models are not merely technical amalgams; they offer valuable insights into how AI is beginning to "think" in a manner reminiscent of the complex, layered thinking found in biological cognition.

Generative AI, alongside XAI, emphasizes this interaction. Generative models, such as GANs and diffusion models, excel at producing content, but their biases and accountability issues often remain unclear. This lack of transparency can lead to hesitance in implementing these models due to trust concerns. By being explicit about the generative process, I, as a stakeholder, can advocate for generative models while also embracing my responsibility for human intelligence and creativity. Since

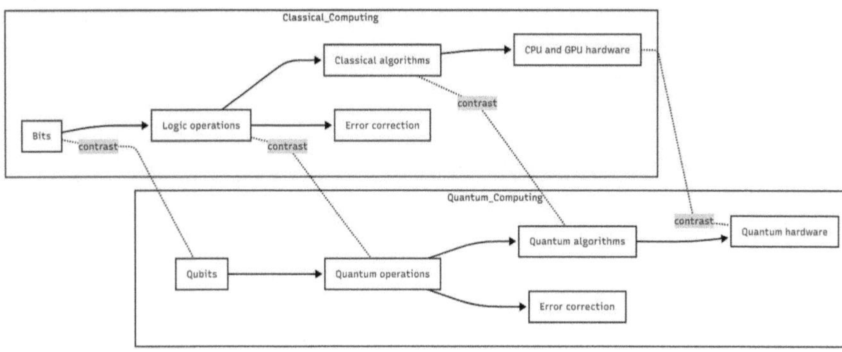


**Fig. 1.49** Classical vs. Quantum Computing

generative models are rooted in historical context—reflecting both biases and past failures—and since XAI relates to the imperative for clear judgment, the resulting synergy fosters a balance between creativity and accountability [211]. This enables AI systems to be evaluated in high-stakes areas such as medicine and law.

Neuromorphic computing further enriches this diverse field by offering bio-inspired hardware capable of real-time, low-power computation. The integration of neuromorphic processors with generative systems enables continuous learning and sensory adaptation, similar to biological processes that facilitate evolutionary capacity and genetic plasticity. This is particularly crucial for robotics and edge devices that require real-time processing and energy efficiency.

One of the most promising hybrid pathways lies at the intersection of Federated Learning (FL) and Quantum Machine Learning (QML). FL aims to decentralize learning while preserving privacy, yet it imposes a computational burden on edge devices, which can limit its effectiveness. Conversely, QML offers exponential speedups for certain learning tasks. Combining FL and QML could lead to a privacy-preserving, quantum-enhanced architecture for distributed learning across various nodes with minimal latency. This model is especially useful for healthcare data, which is often fragmented across national or international institutions but holds significant collective value.

Moreover, it is reasonable to believe that FL, XAI, and QML could reshape the trust landscape for AI in the future. By facilitating decentralized training, providing clear and interpretable models, and harnessing quantum computing (as well as other innovative technologies) for model inference, these systems can address many ethical and technical challenges associated with broad AI usage in heavily regulated environments. The outcome will be not only faster models but also fairer and more responsible ones (Fig. 1.50 and Table 1.16).

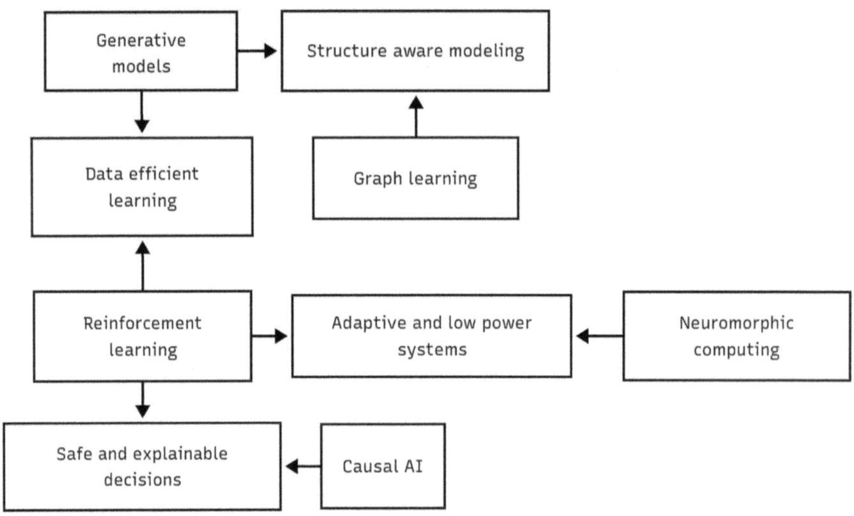


**Fig. 1.50** Synergy Points Between the Five Emerging AI Technologies

**Table 1.16** Strengths and Challenges of Each AI Technology

| AI Technology | Key Strengths | Main Challenges |
|---|---|---|
| Machine learning (traditional) | Well-established algorithms, interpretable models (e.g., linear regression, decision trees), and relatively low computational requirements | Requires careful feature engineering, performance is limited by feature quality, may struggle with unstructured data |
| Deep learning | Automatically learns features from raw data, state-of-the-art results in vision, speech, and language tasks | Data-hungry, high computational cost, limited interpretability, risk of overfitting |
| Reinforcement learning | Learns optimal sequential decisions through interaction, can surpass human performance in games and control problems | Requires a large exploration space and simulations, unstable training, and sparse rewards |
| Natural language processing (NLP) | Enables human-machine communication, supports tasks like translation, summarization, and question answering | Sensitive to bias in training data, requires extensive text corpora, and contextual understanding is still imperfect |
| Computer vision | Accurate image and video recognition, powerful for autonomous driving, medical imaging, and surveillance | Requires large labeled datasets, privacy concerns, vulnerable to adversarial attacks |
| Expert systems/ symbolic AI | Transparent reasoning through explicit rules, good for well-defined domains | Difficult to scale or maintain, brittle when encountering unseen scenarios |
| Generative models (GANs, diffusion) | High-fidelity data generation for images, audio, and text; powerful for augmentation and creative tasks | High computational demand, risk of misuse (deepfakes), training instability (GANs) |
| Graph neural networks (GNNs) | Capture complex relationships in network data (social graphs, molecules) | Scalability to very large graphs, interpretability, and sensitivity to graph noise |
| Neuromorphic computing | Low-power event-driven processing inspired by the brain | Immature hardware ecosystem, programming models still evolving |

The future of AI is likely to include these emerging technologies as essential components of cognitive architectures that physically and functionally reference the human brain. It is possible (and perhaps even probable) that federated neuromorphic infrastructure, explainable generative interfaces, and diagnostic tools powered by quantum computing will evolve into a cohesive ecosystem of learning and reasoning [212]. These systems would learn to solve new problems using minimal data while intentionally ensuring that their outputs reflect human values and understanding.

## References

1. Xu B, Yang G. Interpretability research of deep learning: a literature survey. Inf Fusion. 2025;115:102721.

2. Yang W, Wei Y, Wei H, Chen Y, Huang G, Li X, Kang B. Survey on explainable AI: from approaches, limitations and applications aspects. Human-Centric Intelligent Systems. 2023;3(3):161–88.
3. Colelough BC, Regli W. Neuro-symbolic AI in 2024: a systematic review. arXiv:2501.05435. 2025;
4. Ranaldi L. Survey on the role of mechanistic interpretability in generative AI. Big Data Cogn Comput. 2025;9(8):193.
5. Kazmierczak R, Berthier E, Frehse G, Franchi G. Explainability and vision foundation models: a survey. Inf Fusion. 2025;122:103184.
6. Zhao H, Chen H, Yang F, Liu N, Deng H, Cai H, Du M. Explainability for large language models: a survey. ACM Trans Intell Syst Technol. 2024;15(2):1–38.
7. Michel-Delétie C, Sarker MK. Neuro-symbolic methods for trustworthy AI: a systematic review. Neurosymbolic Artif Intell. 2024;
8. Akter MS, Sultan MF, Cuzzocrea A (2025) Neuro-symbolic methods in natural language processing: a review
9. Gui J, Chen T, Zhang J, Cao Q, Sun Z, Luo H, Tao D. A survey on self-supervised learning: algorithms, applications, and future trends. IEEE Trans Pattern Anal Mach Intell. 2024;46(12):9052–71.
10. Parisi GI, Kemker R, Part JL, Kanan C, Wermter S. Continual lifelong learning with neural networks: a review. Neural Netw. 2019;113:54–71.
11. Kaplan J, McCandlish S, Henighan T, Brown TB, Chess B, Child R, Amodei D. Scaling laws for neural language models. arXiv:2001.08361. 2020;
12. Wei J, Wang X, Schuurmans D, Bosma M, Xia F, Chi E, Zhou D. Chain-of-thought prompting elicits reasoning in large language models. Adv Neural Inf Process Syst. 2022;35:24824–37.
13. Hendrycks D, Burns C, Basart S, Zou A, Mazeika M, Song D, Steinhardt J. Measuring massive multitask language understanding. arXiv:2009.03300. 2020;
14. Lecue F. On the role of knowledge graphs in explainable AI. Semant Web. 2020;11(1):41–51.
15. Bhutani A, Snaria A. The past, present and future in artificial intelligence. GLMS J Manag Rev Transform. 2023;2:132–40.
16. Zhou J, Cui G, Hu S, Zhang Z, Yang C, Liu Z, Sun M. Graph neural networks: a review of methods and applications. AI Open. 2020;1:57–81.
17. Bubeck S, Chandrasekaran V, Eldan R, Gehrke J, Horvitz E, Kamar E, Zhang Y. Sparks of artificial general intelligence: early experiments with GPT-4. arXiv:2303.12712. 2023;
18. Guido R, Ferrisi S, Lofaro D, Conforti D. An overview on the advancements of support vector machine models in healthcare applications: a review. Information. 2024;15(4):235.
19. Halder RK, Uddin MN, Uddin MA, Aryal S, Khraisat A. Enhancing K-nearest neighbor algorithm: a comprehensive review and performance analysis of modifications. J Big Data. 2024;11(1):113.
20. Li J, Guo S, Ma R, He J, Zhang X, Rui D, Ding Y, Li Y, Jian L, Cheng J, Guo H. Comparison of the effects of imputation methods for missing data in predictive modelling of cohort study datasets. BMC Med Res Methodol. 2024;24(1):41.
21. Miyazaki Y, Kawakami M, Kondo K, Hirabe A, Kamimoto T, Akimoto T, Tsuji T. Logistic regression analysis and machine learning for predicting post-stroke gait independence: a retrospective study. Sci Rep. 2024;14(1):21273.
22. Wu T, Wei Y, Wu J, Yi B, Li H. Logistic regression technique is comparable to complex machine learning algorithms in predicting cognitive impairment related to post intensive care syndrome. Sci Rep. 2023;13(1):2485.
23. Harris JK. Primer on binary logistic regression. Fam Med Community Health. 2021;9(Suppl 1):e001290.
24. Cervantes J, Garcia-Lamont F, Rodríguez-Mazahua L, Lopez A. A comprehensive survey on support vector machine classification: applications, challenges and trends. Neurocomputing. 2020;408:189–215.

25. Richardson E, Trevizani R, Greenbaum JA, Carter H, Nielsen M, Peters B. The receiver operating characteristic curve accurately assesses imbalanced datasets. Patterns. 2024;5(6):e02485.
26. Hancock JT, Khoshgoftaar TM, Johnson JM. Evaluating classifier performance with highly imbalanced big data. J Big Data. 2023;10(1):42.
27. McDermott M, Zhang H, Hansen L, Angelotti G, Gallifant J. A closer look at AUROC and AUPRC under class imbalance. Adv Neural Inf Process Syst. 2024;37:44102–63.
28. Chen W, Yang K, Yu Z, Shi Y, Chen CP. A survey on imbalanced learning: latest research, applications and future directions. Artif Intell Rev. 2024;57(6):137.
29. Costa VG, Pedreira CE. Recent advances in decision trees: an updated survey. Artif Intell Rev. 2023;56(5):4765–800.
30. Mohammadi-Pirouz Z, Hajian-Tilaki K, Sadeghi Haddat-Zavareh M, Amoozadeh A, Bahrami S. Development of decision tree classification algorithms in predicting mortality of COVID-19 patients. Int J Emerg Med. 2024;17(1):126.
31. Jones L, Barnett A, Vagenas D. Common misconceptions held by health researchers when interpreting linear regression assumptions: a cross-sectional study. PLoS One. 2025;20(6):e0299617.
32. Jones L, Barnett A, Vagenas D. Linear regression reporting practices for health researchers: a cross-sectional meta-research study. PLoS One. 2025;20(3):e0305150.
33. Mienye ID, Jere N. A survey of decision trees: concepts, algorithms, and applications. IEEE Access. 2024;12:86716–27.
34. Abedinia A, Seydi V. Building semi-supervised decision trees with semi-CART algorithm. Int J Mach Learn Cybern. 2024;15(10):4493–510.
35. Lu J, Lu X, Wang Y, Zhang H, Han L, Zhu B, Wang B. Comparison between logistic regression and machine learning algorithms on prediction of noise-induced hearing loss and investigation of SNP loci. Sci Rep. 2025;15(1):15361.
36. Bist US, Singh N. Analysis of recent advancements in support vector machine. Concurr Comput Pract Exp. 2022;34(25):e7270.
37. Jo N, Aghaei S, Benson J, Gomez A, Vayanos P. Learning optimal fair decision trees: tradeoffs between interpretability, fairness, and accuracy. In: Proceedings of the eighth AAAI/ACM conference on AI, Ethics, and Society (AIES-25); 2023. p. 181–92.
38. Nanfack G, Temple P, Frénay B. Constraint enforcement on decision trees: a survey. ACM Comput Surv. 2022;54(10s):1–36.
39. Ruder S, Peters ME, Swayamdipta S, Wolf T. Transfer learning in natural language processing. In: Proceedings of NAACL tutorials; 2019. p. 15–8.
40. Zhou ZH. Ensemble methods: foundations and algorithms. CRC Press; 2025.
41. Kunapuli G. Ensemble methods for machine learning. Simon & Schuster; 2023.
42. Agrawal P. A survey on hyperparameter optimization of machine learning models. In: Proceedings of 2nd international conference on disruptive technologies (ICDT 2024); 2024. p. 11–5.
43. Plested J, Gedeon T. Deep transfer learning for image classification: a survey. arXiv:2205.09904. 2022;
44. Yu F, Xiu X, Li Y. A survey on deep transfer learning and beyond. Mathematics. 2022;10(19):3619.
45. Farmanesh A, Sanchis RG, Ordieres-Meré J. Comparison of deep transfer learning against contrastive learning in industrial quality applications for heavily unbalanced data scenarios when data augmentation is limited. Sensors (Basel). 2025;25(10):3048.
46. Cao Y, Xiang H, Zhang H, Zhu Y, Ting KM. Anomaly detection based on isolation mechanisms: a survey. Mach Intell Res. 2025:1–17.
47. Chua W, Pajas ALD, Castro CS, Panganiban SP, Pasuquin AJ, Purganan MJ, Malupeng R, Pingad DJ, Orolfo JP, Lua HH, Velasco LC. Web traffic anomaly detection using isolation forest. Informatics. 2024;11(4):83.

48. Badnjevic A, Deumic A, Trakic A, Pokvic LG. A novel method for conformity assessment testing of mechanical ventilators for post-market surveillance purposes. Technol Health Care. 2023;31(1):367–76.
49. Arcudi A, Frizzo D, Masiero C, Susto GA. Enhancing interpretability and generalizability in extended isolation forests. Eng Appl Artif Intell. 2024;138:109409.
50. Yepmo V, Smits G, Lesot MJ, Pivert O. Leveraging an isolation forest to anomaly detection and data clustering. Data Knowl Eng. 2024;151:102302.
51. Gałka Ł. Optimized deep isolation forest. Pattern Recogn Lett. 2025;197:88–94.
52. Monemizadeh V, Kiani K. Detecting anomalies using rotated isolation forest. Data Min Knowl Discov. 2025;39(3):1–31.
53. Leveni F, Magri L, Alippi C, Boracchi G. Preference isolation forest for structure-based anomaly detection. arXiv:2505.10876. 2025;
54. Binetti MS, Uricchio VF, Massarelli C. Isolation forest for environmental monitoring: a data-driven approach to land management. Environments. 2025;12(4):116.
55. Badnjevic A, Deumic A, Dzemic Z, Pokvic LG. A novel method for conformity assessment testing of anaesthesia machines for post-market surveillance purposes. Technol Health Care. 2023;31(1):377–87.
56. Eremin EO. Unsupervised anomaly detection on cybersecurity data streams: a case with BETH dataset. Int J Open Inf Technol. 2025;13(6):107–13.
57. Agyemang EF. Anomaly detection using unsupervised machine learning algorithms: a simulation study. Sci Afr. 2024;26:e02386.
58. Seman LO, Aquino LS, Stefenon SF, Yow KC, Mariani VC, Coelho LS. Simultaneously anomaly detection and forecasting for predictive maintenance using a zero-cost differentiable architecture search-based network. Comput Ind Eng. 2025:111412.
59. Magjarević R, Badnjevic A. Inspection and testing of electrocardiographs (ECG) devices. In: Inspection of medical devices. Series in biomedical engineering. Cham: Springer; 2024.
60. Yang Y, Liu S, Yang Q. Background subspace projection-based Mahalanobis distance method for hyperspectral anomaly detection. In: Proceedings of 7th international conference on artificial intelligence and pattern recognition; 2024. p. 594–600.
61. Astivia OLO. A method to simulate multivariate outliers with known Mahalanobis distances for normal and non-normal data. Methods Psychol. 2024;11:100157.
62. Lin J, Xiong K. Mahalanobis distance-based robust approaches against false data injection attacks on dynamic power state estimation. Comput Secur. 2021;108:102326.
63. Mozharovskyi P, Valla R. Anomaly detection using data depth: multivariate case. Int J Data Sci Anal. 2025:1–26.
64. Badnjevic A, Magjarevic R, Mrdjanovic E, Pokvic LG. A novel method for conformity assessment testing of electrocardiographs for post-market surveillance purposes. Technol Health Care. 2023;31(1):307–15.
65. Pinheiro C, Guerreiro S, Mamede HS. A survey on association rule mining for enterprise architecture model discovery. Bus Inf Syst Eng. 2023;66(6):777–98.
66. Geng X, Yang Z, Jiao L, Zhou ZJ, Ma Z. Association rule-based classification: a comprehensive review of methodologies and applications. Expert Syst Appl. 2025:127454.
67. Badnjević A, Zebić EZ. Bosnia and Herzegovina's e-identity initiative: a technological leap in digital solutions and governance. Technol Health Care. 2024;32(5):2853–8.
68. Cheng WS, Lin YT, Huang PY, Chen JC, Lin KW. A fast and highly scalable frequent pattern mining algorithm. Futur Gener Comput Syst. 2024;160:854–68.
69. Hunyadi ID, Constantinescu N, Ţicleanu O-A. Efficient discovery of association rules in e-commerce: comparing candidate generation and pattern growth techniques. Appl Sci. 2025;15(10):5498.
70. Kallay P, Mihoc TD. Comparative analysis of frequent pattern mining algorithms. Acta Univ Sapientiae Inform. 2025;17(1):8.

71. Li W. Improvements to the Apriori algorithm and its application in educational decision-making systems. In: Proceedings of 2nd International Conference on Generative Artificial Intelligence and Information Security; 2025. p. 367–73.
72. He M, Chen X. Anomaly detection algorithm for big data based on isolation forest algorithm. J Comput Methods Sci Eng. 2025:14727978251337984.
73. Khan IA, Chen H-Y, Sharma S, Sharma C. Recursive queried frequent patterns algorithm: determining frequent pattern sets from database. Information. 2025;16(9):746.
74. Yang Y, Zhao J, Xu X, Cao K, Yuan S, Xie L. Unsupervised anomaly detection for autonomous robots via Mahalanobis SVDD with audio-IMU fusion. arXiv:2505.05811. 2025;
75. Amin M, Efendi S, Candra A. Performance improvement of Apriori algorithm with transaction reduction and hash based approach. In: Proceedings of 4th international conference of science and information technology in smart administration (ICSINTESA); 2024. p. 517–22.
76. Badnjevic A, Deumic A, Biskovic D, Pokvic LG. A novel method for conformity assessment testing of dialysis machines for post-market surveillance purposes. Technol Health Care. 2023;31(1):357–65.
77. Lin J, Fukuyama J. Calibrating dimension reduction hyperparameters in the presence of noise. PLoS Comput Biol. 2024;20(9):e1012427.
78. Ikotun AM, Habyarimana F, Ezugwu AE. Benchmarking validity indices for evolutionary K-means clustering performance. Sci Rep. 2025;15(1):21842.
79. Zhou S, Xu H, Zheng Z, Chen J, Li Z, Bu J, Ester M. A comprehensive survey on deep clustering: taxonomy, challenges, and future directions. ACM Comput Surv. 2024;57(3):1–38.
80. Martinović M, Kosović M, Spahić L, Softić A, Pokvić LG, Badnjević A. Machine learning for improved medical device management: a focus on dialysis machines. Technol Health Care. 2025; [early online]: 9287329251328815.
81. Sheng G, Wang Q, Pei C, Gao Q. Contrastive deep embedded clustering. Neurocomputing. 2022;514:13–20.
82. Liu J, Vinck M. Improved visualization of high-dimensional data using the distance-of-distance transformation. PLoS Comput Biol. 2022;18(12):e1010764.
83. Nadjafikhah M, Nasiri M. A comparative study of manifold learning methods for scRNA-seq with a trajectory-aware metric. Sci Rep. 2025;15(1):28923.
84. Badnjević A, Pokvić LG, Smajlhodžić-Deljo M, et al. Application of artificial intelligence for the classification of the clinical outcome and therapy in patients with viral infections: the case of COVID-19. Technol Health Care. 2024;32(3):1859–70.
85. Trozzi F, Wang X, Tao P. UMAP as a dimensionality reduction tool for molecular dynamics simulations of biomacromolecules: a comparison study. J Phys Chem B. 2021;125(19):5022–34.
86. Lázaro C, Angulo C. Using UMAP for partially synthetic healthcare tabular data generation and validation. Sensors. 2024;24(23):7843.
87. Meilă M, Zhang H. Manifold learning: what, how, and why. Annu Rev Stat Its Appl. 2024;11(1):393–417.
88. Yousuff M, Babu R, Rathinam A. Nonlinear dimensionality reduction based visualization of single-cell RNA sequencing data. J Anal Sci Technol. 2024;15(1):1.
89. Berahmand K, Daneshfar F, Salehi ES, Li Y, Xu Y. Autoencoders and their applications in machine learning: a survey. Artif Intell Rev. 2024;57(2):28.
90. Asperti A, Evangelista D, Loli Piccolomini E. A survey on variational autoencoders from a green AI perspective. SN Comput Sci. 2021;2(4):301.
91. Ma H, Wu W. A deep clustering framework integrating pairwise constraints and a VMF mixture model. Electron Res Arch. 2024;32(6)
92. Ju W, Fang Z, Gu G, Liu Z, Long Q, Qiao Z, Zhang M. A comprehensive survey on deep graph representation learning. Neural Netw. 2024;173:106207.
93. Lossio-Ventura JA, Gonzales S, Morzan J, Alatrista-Salas H, Hernandez-Boussard T, Bian J. Evaluation of clustering and topic modeling methods over health-related tweets and emails. Artif Intell Med. 2021;117:102096.

94. Xue G, Zhong M, Li J, Chen J, Zhai C, Kong R. Dynamic network embedding survey. Neurocomputing. 2022;472:212–23.
95. Badnjevic A, Deumic A, Smajlhodzic-Deljo M, Pokvic LG. A novel method for conformity assessment testing of infusion and perfusion pumps for post-market surveillance purposes. Technol Health Care. 2023;31(1):347–55.
96. Badnjevic A, Deumic A, Ademovic A, Pokvic LG. A novel method for conformity assessment testing of therapeutic ultrasounds for post-market surveillance purposes. Technol Health Care. 2023;31(1):339–46.
97. Badnjevic A. Evidence-based maintenance of medical devices: current shortage and pathway towards solution. Technol Health Care. 2023;31(1):293–305.
98. Badnjević A, Pokvić LG, Deumić A, Bećirović LS. Post-market surveillance of medical devices: a review. Technol Health Care. 2022;30(6):1315–29.
99. Mujkić E, Baralić A, Ombašić L, Bećirović LS, Pokvić LG, Badnjević A. Machine intelligence in biomedical data modeling, processing, and analysis. In: Proceedings of 11th Mediterranean conference on embedded computing (MECO); 2022. p. 1–10.
100. Licen S, Astel A, Tsakovski S. Self-organizing map algorithm for assessing spatial and temporal patterns of pollutants in environmental compartments: a review. Sci Total Environ. 2023;878:163084.
101. Guérin A, Chauvet P, Saubion F. A survey on recent advances in self-organizing maps. arXiv:2501.08416. 2024;
102. Fan X, Zhang S, Xue X, Jiang R, Fan S, Kou H. An improved self-organizing map (SOM) based on virtual winning neurons. Symmetry. 2025;17(3):449.
103. Chari T, Pachter L. The specious art of single-cell genomics. PLoS Comput Biol. 2023;19(8):e1011288.
104. Mittal M, Gujjar P, Prasad G, Devadas RM, Ambreen L, Kumar V. Dimensionality reduction using UMAP and t-SNE technique. In: Proceedings of 2nd international conference on advanced intelligent technologies (ICAIT); 2024. p. 1–5.
105. Badnjevic A, Spahic L, Jordamovic NB, Pokvic LG. A novel method for conformity assessment testing of infant incubators for post-market surveillance purposes. Technol Health Care. 2023;31(1):389–99.
106. Hassan BA, Tayfor NB, Hassan AA, Ahmed AM, Rashid TA, Abdalla NN. From A-to-Z review of clustering validation indices. Neurocomputing. 2024;601:128198.
107. Liang S, Pan Z, Liu W, Yin J, De Rijke M. A survey on variational autoencoders in recommender systems. ACM Comput Surv. 2024;56(10):1–40.
108. Touvron H, Bojanowski P, Caron M, Cord M, El-Nouby A, Grave E, Jégou H. ResMLP: feedforward networks for image classification with data-efficient training. IEEE Trans Pattern Anal Mach Intell. 2022;45(4):5314–21.
109. Hou Q, Jiang Z, Yuan L, Cheng MM, Yan S, Feng J. Vision permutator: a permutable MLP-like architecture for visual recognition. IEEE Trans Pattern Anal Mach Intell. 2022;45(1):1328–34.
110. Yu T, Li X, Cai Y, Sun M, Li P. S2-MLP: spatial-shift MLP architecture for vision. In: Proceedings of IEEE/CVF winter conference on applications of computer vision (WACV); 2022. p. 297–306.
111. Lee-Thorp J, Ainslie J, Eckstein I, Ontanon S. FNet: mixing tokens with Fourier transforms. arXiv:2105.03824. 2021;
112. Dubey SR, Singh SK, Chaudhuri BB. Activation functions in deep learning: a comprehensive survey and benchmark. Neurocomputing. 2022;503:92–108.
113. Kunc V, Kléma J. Three decades of activations: a comprehensive survey of 400 activation functions for neural networks. arXiv:2402.09092. 2024;
114. Somvanshi S, Das S, Javed SA, Antariksa G, Hossain A. A survey on deep tabular learning. arXiv:2410.12034. 2024;
115. Saleem R, Yuan B, Kurugollu F, Anjum A, Liu L. Explaining deep neural networks: a survey on the global interpretation methods. Neurocomputing. 2022;513:165–80.

116. Ying H, Song M, Tang Y, Xiao S, Xiao Z. Enhancing deep neural network training efficiency and performance through linear prediction. Sci Rep. 2024;14(1):15197.
117. Spahić L, Pokvić LG, Badnjević A. Inspection and testing of infusion pumps. In: Inspection of medical devices, Series in biomedical engineering. Cham: Springer; 2024.
118. Foret P, Kleiner A, Mobahi H, Neyshabur B. Sharpness-aware minimization for efficiently improving generalization. arXiv:2010.01412. 2020;
119. Tolstikhin IO, Houlsby N, Kolesnikov A, Beyer L, Zhai X, Unterthiner T, Dosovitskiy A. MLP-mixer: an all-MLP architecture for vision. Adv Neural Inf Process Syst. 2021;34:24261–72.
120. Chen X, Liang C, Huang D, Real E, Wang K, Pham H, Le QV. Symbolic discovery of optimization algorithms. Adv Neural Inf Process Syst. 2023;36:49205–33.
121. Chen L, Liu B, Liang K, Liu Q. Lion secretly solves constrained optimization: as Lyapunov predicts. arXiv:2310.05898. 2023;
122. Augustine MT. A survey on universal approximation theorems. arXiv:2407.12895. 2024;
123. Pillonetto G, Aravkin A, Gedon D, Ljung L, Ribeiro AH, Schön TB. Deep networks for system identification: a survey. Automatica. 2025;171:111907.
124. Wu L, Li J, Wang Y, Meng Q, Qin T, Chen W, Liu TY. R-drop: regularized dropout for neural networks. Adv Neural Inf Process Syst. 2021;34:10890–905.
125. Salehin I, Kang D-K. A review on dropout regularization approaches for deep neural networks within the scholarly domain. Electronics. 2023;12(14):3106.
126. Dereich S, Jentzen A, Riekert A. Averaged Adam accelerates stochastic optimization in the training of deep neural network approximations for partial differential equation and optimal control problems. arXiv:2501.06081. 2025;
127. Cheng H, Zhang M, Shi JQ. A survey on deep neural network pruning: taxonomy, comparison, analysis, and recommendations. IEEE Trans Pattern Anal Mach Intell. 2024;46:10558–78.
128. Dosovitskiy A, Beyer L, Kolesnikov A, Weissenborn D, Zhai X, Unterthiner T, Houlsby N. An image is worth 16x16 words: transformers for image recognition at scale. arXiv:2010.11929. 2020;
129. Touvron H, Cord M, Douze D, Massa F, Sablayrolles A, Jégou H. Training data-efficient image transformers & distillation through attention. Proc Int Conf Mach Learn. 2021:10347–57.
130. Liu Z, Lin Y, Cao Y, Hu H, Wei Y, Zhang Z, Guo B. Swin transformer: hierarchical vision transformer using shifted windows. In: Proceedings of the IEEE international conference on computer vision; 2021. p. 10012–22.
131. Bećirović F, Spahić L, Merdović N, Gurbeta Pokvić L, Badnjević A. Advancement of post-market surveillance of medical devices leveraging artificial intelligence: patient monitors case study. Technol Health Care. 2024;33(2):974–80.
132. Liu Z, Mao H, Wu CY, Feichtenhofer C, Darrell T, Xie S. A ConvNet for the 2020s. In: Proceedings of IEEE/CVF conference on computer vision and pattern recognition (CVPR); 2022. p. 11976–86.
133. Tan M, Le Q. EfficientNetV2: smaller models and faster training. In: Proceedings of international conference on machine learning; 2021. p. 10096–106.
134. Ding X, Zhang X, Ma N, Han J, Ding G, Sun J. RepVGG: making VGG-style convnets great again. In: Proceedings of IEEE/CVF conference on computer vision and pattern recognition; 2021. p. 13733–42.
135. Han K, Wang Y, Tian Q, Guo J, Xu C, Xu C. GhostNet: more features from cheap operations. In: Proceedings of IEEE/CVF conference on computer vision and pattern recognition; 2020. p. 1580–9.
136. Medjedovic E, Stanojevic M, Jonuzovic-Prosic S, et al. Artificial intelligence as a new answer to old challenges in maternal-fetal medicine and obstetrics. Technol Health Care. 2024;32(3):1273–87.
137. Zhang H, Wu C, Zhang Z, Zhu Y, Lin H, Zhang Z, Smola A. ResNeSt: split-attention networks. In: Proceedings of IEEE/CVF conference on computer vision and pattern recognition (CVPR); 2022. p. 2736–46.

138. Chen T, Kornblith S, Norouzi M, Hinton G. A simple framework for contrastive learning of visual representations. In: Proceedings of international conference on machine learning; 2020. p. 1597–607.
139. Chen X, Fan H, Girshick R, He K. Improved baselines with momentum contrastive learning. arXiv:2003.04297. 2020;
140. Kolesnikov A, Beyer L, Zhai X, Puigcerver J, Yung J, Gelly S, Houlsby N. Big transfer (BiT): general visual representation learning. In: Proceedings of European conference on computer vision; 2020. p. 491–507.
141. Beltagy I, Peters ME, Cohan A. Longformer: the long-document transformer. arXiv:2004.05150. 2020;
142. Zaheer M, Guruganesh G, Dubey KA, Ainslie J, Alberti C, Ontanon S, Ahmed A. Big Bird: Transformers for longer sequences. Adv Neural Inf Process Syst. 2020;33:17283–97.
143. Khan S, Naseer M, Hayat M, Zamir SW, Khan FS, Shah M. Transformers in vision: a survey. ACM Comput Surv. 2022;54(10s):1–41.
144. Raffel R, Shazeer N, Roberts A, Lee K, Narang S, Matena M, Liu PJ. Exploring the limits of transfer learning with a unified text-to-text transformer. J Mach Learn Res. 2020;21(140):1–67.
145. Spahić L, Sredović U, Kurpejović Z, Mrdanović E, Pokvić G, Badnjević A. Machine learning for improved medical device management: a focus on infant incubators. Technol Health Care. 2025;33:2034–40.
146. Wu Z, Pan S, Chen F, Long G, Zhang C, Yu PS. A comprehensive survey on graph neural networks. IEEE Trans Neural Netw Learn Syst. 2020;32(1):4–24.
147. Hu W, Fey M, Zitnik M, Dong Y, Ren H, Liu B, Leskovec J. Open graph benchmark: datasets for machine learning on graphs. Adv Neural Inf Process Syst. 2020;33:22118–33.
148. Bronstein MM, Bruna J, Cohen T, Veličković P. Geometric deep learning: grids, groups, graphs, geodesics, and gauges. arXiv:2104.13478. 2021;
149. Hewamalage H, Bergmeir C, Bandara K. Recurrent neural networks for time series forecasting: current status and future directions. Int J Forecast. 2021;37(1):388–427.
150. Hafizović L, Čaušević A, Deumić A, Bećirović LS, Pokvić LG, Badnjević A. The use of artificial intelligence in diagnostic medical imaging: systematic literature review. In: Proceedings of IEEE 21st international conference on bioinformatics and bioengineering (BIBE); 2021. p. 1–6.
151. Bommasani R. On the opportunities and risks of foundation models. arXiv:2108.07258. 2021;
152. Hüllermeier E, Waegeman W. Aleatoric and epistemic uncertainty in machine learning: an introduction to concepts and methods. Mach Learn. 2021;110(3):457–506.
153. Praharaj M, Sain D, Mohan BM. Modelling and analysis of an interval type-2 Mamdani fuzzy PID controller. IFAC-PapersOnLine. 2022;55(1):728–33.
154. Pokvić LG, Deumić A, Softić A, Badnjević A. Application of artificial intelligence for management of maintenance and prediction of performances. In: Inspection of medical devices. Series in biomedical engineering. Cham: Springer; 2024.
155. Moral A, Castiello C, Magdalena L, Mencar C. Explainable fuzzy systems. New York: Springer Int Publ; 2021.
156. Zhao L, Lin Z, Sun R, Wang A. A review of state-of-the-art methodologies and applications in action recognition. Electronics. 2024;13(23):4733.
157. Liebowitz J. Expert systems with applications. Expert Syst Appl. 2013;40
158. Sharma YK, Ahmed G, Saini DK. Comparative performance analysis of Mamdani and Sugeno fuzzy inference systems for sustainable cluster formation in WSNs. J Intell Fuzzy Syst. 2024:18758967251340448.
159. Türk İ, Kılıç H, Haydaroğlu C, Top A. Robust load frequency control in hybrid microgrids using type-3 fuzzy logic under stochastic variations. Symmetry. 2025;17(6):853.
160. Spahić L, Jeremić L, Lalatović I, et al. Machine learning for improved medical device management: a focus on defibrillator performance. Technol Health Care. 2024;33(2):737–43.

161. Abdrakhmanov R, Berkimbayev K, Seitmuratov A, Ibashova A, Aliyeva A, Nurmukhanbetova G. Intelligent fuzzy-PID temperature control system for ensuring comfortable microclimate in an intelligent building. Int J Adv Comput Sci Appl. 2024;15(3)
162. Yakubu AU, Xiong S, Jiang Q, Zhao J, Wu Z, Wang H, Wangsen H. Fuzzy-based thermal management control analysis of vehicle air conditioning system. Int J Hydrogen Energy. 2024;77:834–43.
163. Safdari M, Al Janaideh M, Siddiqui K, Aliabadi AA. Weather-adaptive fuzzy control of set-points for energy-efficient HVAC in urban buildings. J Build Eng. 2025;104:112317.
164. Nishanth FP, Dash SK, Mahapatro SR. Critical study of type-2 fuzzy logic control from theory to applications: a state-of-the-art comprehensive survey. e-Prime Adv Electr Eng Electron Energ. 2024;10:100771.
165. Badnjevic A, Deumic A, Softic A, Pokvic LG. A novel method for conformity assessment testing of patient monitors for post-market surveillance purposes. Technol Health Care. 2023;31(1):327–37.
166. Mittal K, Jain A, Vaisla KS, Castillo O, Kacprzyk J. A comprehensive review on type-2 fuzzy logic applications: past, present and future. Eng Appl Artif Intell. 2020;95:103916.
167. Alcantud JCR. Ranked hesitant fuzzy sets for multi-criteria multi-agent decisions. Expert Syst Appl. 2022;209:118276.
168. Haighton R, Asgharnia A, Schwartz H, Givigi S. An adaptable fuzzy reinforcement learning method for non-stationary environments. Neurocomputing. 2024;604:128309.
169. Małolepsza O, Mikołajewski D, Prokopowicz P. Using fuzzy logic to analyse weather conditions. Electronics. 2024;14(1):85.
170. Castillo O, Melin P. A review on interval type-2 fuzzy logic applications in intelligent control. Inf Sci. 2014;279:615–31.
171. Ho J, Jain A, Abbeel P. Denoising diffusion probabilistic models. Adv Neural Inf Process Syst. 2020;33:6840–51.
172. Dhariwal P, Nichol A. Diffusion models beat GANs on image synthesis. Adv Neural Inf Process Syst. 2021;34:8780–94.
173. Rombach R, Blattmann A, Lorenz D, Esser P, Ommer B. High-resolution image synthesis with latent diffusion models. In: Proceedings of the IEEE/CVF conference on computer vision and pattern recognition (CVPR); 2022. p. 10684–95.
174. Ramesh A, Dhariwal P, Nichol A, Chu C, Chen M. Hierarchical text-conditional image generation with CLIP latents. arXiv:2204.06125. 2022;1(2):3.
175. Saharia C, Chan W, Saxena S, Li L, Whang J, Denton EL, Norouzi M. Photorealistic text-to-image diffusion models with deep language understanding. Adv Neural Inf Process Syst. 2022;35:36479–94.
176. Kairouz P, McMahan HB, Avent B, Bellet A, Bennis M, Bhagoji AN, Zhao S. Advances and open problems in federated learning. Foundations and Trends® in Machine Learning. 2021;14(1–2):1–210.
177. Merdović N, Spahić L, Hundur M, Pokvić LG, Badnjević A. Advancement of post-market surveillance of medical devices leveraging artificial intelligence: infusion pumps case study. Technol Health Care. 2024;33(2):915–21.
178. Nguyen DC, Ding M, Pathirana PN, Seneviratne A, Li J, Poor HV. Federated learning for internet of things: a comprehensive survey. IEEE Commun Surv Tutor. 2021;23(3):1622–58.
179. Li Q, Wen Z, Wu Z, Hu S, Wang N, Li Y, He B. A survey on federated learning systems: vision, hype and reality for data privacy and protection. IEEE Trans Knowl Data Eng. 2021;35(4):3347–66.
180. Bagdasaryan E, Veit A, Hua Y, Estrin D, Shmatikov V. How to backdoor federated learning. In: Proceedings of international conference on artificial intelligence and statistics; 2020. p. 2938–48.
181. Badnjević A, Pokvić LG, Džemić Z, et al. Risks of emergency use authorizations for medical products during outbreak situations: a COVID-19 case study. Biomed Eng Online. 2020;19:75.

182. Cao X, Fang M, Liu J, Gong NZ. FLTrust: Byzantine-robust federated learning via trust bootstrapping. arXiv:2012.13995. 2020;
183. Arrieta AB, Díaz-Rodríguez N, Del Ser J, Bennetot A, Tabik S, Barbado A, Herrera F. Explainable artificial intelligence (XAI): concepts, taxonomies, opportunities and challenges toward responsible AI. Inf Fusion. 2020;58:82–115.
184. Lundberg SM, Erion G, Chen H, DeGrave A, Prutkin JM, Nair B, Lee SI. From local explanations to global understanding with explainable AI for trees. Nat Mach Intell. 2020;2(1):56–67.
185. Mundici D. De Finetti for mathematics undergraduates. arXiv:2107.00250. 2021;
186. Nunes JD, Carvalho M, Carneiro D, Cardoso JS. Spiking neural networks: a survey. IEEE Access. 2022;10:60738–64.
187. Zhang D, Jia S, Wang Q. Recent advances and new frontiers in spiking neural networks. arXiv:2204.07050. 2022;
188. Cerezo M, Arrasmith A, Babbush R, Benjamin SC, Endo S, Fujii K, Coles PJ. Variational quantum algorithms. Nat Rev Phys. 2021;3(9):625–44.
189. Huang HY, Broughton M, Mohseni M, Babbush R, Boixo S, Neven H, McClean JR. Power of data in quantum machine learning. Nat Commun. 2021;12(1):2631.
190. Schuld M. Supervised quantum machine learning models are kernel methods. arXiv:2101.11020. 2021;
191. Mari A, Bromley TR, Izaac J, Schuld M, Killoran N. Transfer learning in hybrid classical–quantum neural networks. arXiv:1912.08278. 2019;
192. Peruničić Ž, Lalatović I, Spahić L, Ašić A, Pokvić LG, Badnjević A. Enhancing mechanical ventilator reliability through machine learning based predictive maintenance. Technol Health Care. 2024;33(3):1288–97.
193. Hundur M, Spahić L, Bećirović F, Gurbeta Pokvić L, Badnjević A. Advancement of post-market surveillance of medical devices leveraging artificial intelligence: ECG devices case study. Technol Health Care. 2025;33:1818–26.
194. Vuković D, Šumar S, Dupovac S, Badnjević A. Cost effectiveness analysis of medical devices in legal metrology system. In: Inspection of medical devices, Series in biomedical engineering. Cham: Springer; 2024.
195. Badnjevic A, Deumic A, Imamovic E, Pokvic LG. A novel method for conformity assessment testing of defibrillators for post-market surveillance purposes. Technol Health Care. 2023;31(1):317–25.
196. Júnior SS, Mendes J, Souza F, Premebida C. Survey on deep fuzzy systems in regression applications: a view on interpretability. Int J Fuzzy Syst. 2023;25(7):2568–89.
197. Valdez F, Castillo O, Melin P. A bibliometric review of type-3 fuzzy logic applications. Mathematics. 2025;13(3):375.
198. Nishanth FP, Dash SK, Mahapatro SR. Critical study of type-2 fuzzy logic control from theory to applications: a state-of-the-art comprehensive survey. e-Prime Adv Electr Eng Electron Energy. 2024;10:100771.
199. Castillo O, Valdez F, Melin P, Ding W. A survey on type-3 fuzzy logic systems and their control applications. IEEE/CAA J Autom Sinica. 2024;11(8):1744–56.
200. Alateeq M, Pedrycz W. Logic-oriented fuzzy neural networks: a survey. Expert Syst Appl. 2024;257:125120.
201. Tang HH, Ahmad NS. Fuzzy logic approach for controlling uncertain and nonlinear systems: a comprehensive review of applications and advances. Syst Sci Control Eng. 2024;12(1):2394429.
202. Apiecionek L. Fuzzy neural networks—a review with case study. Appl Sci. 2025;15(13):6980.
203. Abdalla TY, Abdalla AY, Chyaid AM. Type-3 fuzzy system-based intelligent control approaches and applications. Adv Fuzzy Syst. 2025;2025(1):6661495.
204. Małolepsza O, Mikołajewski D, Prokopowicz P. Adaptation of fuzzy systems based on ordered fuzzy numbers: a review of applications and development prospects. Electronics. 2025;14(12):2341.

205. Zheng Y, Xu Z, Wu T, Yi Z. A systematic survey of fuzzy deep learning for uncertain medical data. Artif Intell Rev. 2024;57(9):230.
206. Paulraj SSS, Deepa T. Energy-efficient data routing using neuro-fuzzy based data routing mechanism for IoT-enabled WSNs. Sci Rep. 2024;14(1):30081.
207. Chang WJ, Lin YH, Ku CC. A comprehensive survey on advanced control techniques for T-S fuzzy systems subject to control input and system output requirements. Processes. 2025;13(3):792.
208. Nguyen TTT, Na J, Nguyen LT, Wang X. Neuro-fuzzy network-based nonlinear hybrid active noise control systems. Entropy. 2025;27(2):138.
209. Sun L, Xia J, She D, Ding W, Jiang J, Liu B, Zhao F. A predictive fuzzy logic and rule-based control approach for practical real-time operation of urban stormwater storage system. Water Res. 2024;266:122437.
210. Shabu SJ, Refonaa J, Mallik S, Dhamodaran D, Grace LJ, Ksibi A, Alshalali TAN. An improved adaptive neuro-fuzzy inference framework for lung cancer detection and prediction on Internet of Medical Things platform. Int J Comput Intell Syst. 2024;17(1):228.
211. Nazha HM, Youssef AM, Darwich MA, Ibrahim TA, Homsieh HE. A comparative study on fuzzy logic-based liquid level control systems with integrated industrial communication technology. Computation. 2025;13(3):60.
212. Dubchak L, Sachenko A, Bodyanskiy Y, Wolff C, Vasylkiv N, Brukhanskyi R, Kochan V. Adaptive neuro-fuzzy system for detection of wind turbine blade defects. Energies. 2024;17(24):6456.

# Chapter 2
# Application of Intelligent Systems in Biomedicine

This chapter demonstrates how intelligent systems are applied to real-world biomedical and healthcare challenges. It begins with AI in medical imaging and diagnostics, describing imaging modalities, preprocessing, deep learning models, clinical validation, and system integration (Sect. 2.1). Intelligent methods for evaluating and maintaining medical device performance are then presented, highlighting predictive maintenance and regulatory compliance (Sect. 2.2). Diagnostic support systems and clinical decision-making frameworks follow, showing how machine learning enhances rare disease diagnosis and real-time clinical guidance (Sect. 2.3). AI-driven personalized medicine is explored through patient stratification, genomics and transcriptomics integration, drug repurposing, virtual screening, predictive treatment modeling, and digital twin development (Sect. 2.4). Wearable biosensors and remote health monitoring are discussed next, focusing on sensor types, signal acquisition, anomaly detection, and edge analytics for chronic disease management (Sect. 2.5). Robotics and AI in surgery and rehabilitation are then examined, covering robot-assisted surgery, surgical planning, exoskeletons, brain–computer interfaces, and rehabilitation robotics (Sect. 2.6). The chapter concludes with ethical, legal, and regulatory considerations, addressing AI training practices, data privacy, bias mitigation, transparency, and compliance with international standards such as FDA and EMA guidelines (Sect. 2.7).

## 2.1 AI in Medical Imaging and Diagnostics

Digital computers have been designed to perform extremely complex tasks, such as proving mathematical theorems and playing chess, with remarkable skill since their invention in the 1940s. Despite ongoing improvements in processing speed and memory capacity, no software can yet fully replicate human adaptability across a



A. Badnjević, L. Spahić, *Intelligent Systems in Biomedicine*, Series in BioEngineering, https://doi.org/10.1007/978-3-032-14785-1_2

wider range of areas or in tasks that require significant common knowledge. However, some programs have reached performance levels comparable to human experts in specific tasks. Consequently, artificial intelligence (AI) is present in various applications, including chatbots, medical diagnosis, search engines, voice recognition, handwriting recognition, and many others.

AI technologies can perform a diverse array of tasks and generate a wide range of outputs, making it difficult to define artificial intelligence with a single definition. Artificial intelligence encompasses computer systems capable of tackling complex activities typically performed by humans, such as thinking, decision-making, producing, learning, understanding language, recognizing patterns, and interacting with their environment [1]. Any artificial system that can perform tasks in various uncertain situations with minimal human oversight or that can learn from experience and improve its performance with new data sets is considered to be exhibiting artificial intelligence.

Machine Learning (ML) and Deep Learning (DL) are two of the numerous subfields within the broader field of artificial intelligence. The goal of machine learning, a branch of artificial intelligence, is to create algorithms that can learn from data and use that knowledge to make decisions or predictions. Similar to humans, machine learning allows machines to learn from experience. There are three main types of machine learning algorithms: supervised learning, unsupervised learning, and reinforcement learning. In supervised learning, a machine learning model is trained using labeled data, where each input sample is associated with a specific output [2]. The model then gains the ability to predict results for unknown data. Unsupervised learning involves training a model on unlabeled data, where the algorithm seeks patterns and structures in the data without predefined outputs. Reinforcement learning is a process in which a system learns to make decisions based on feedback from its environment, such as rewards or penalties.

**Deep Learning**, a subset of machine learning, aims to create artificial neural networks that can learn from vast amounts of data. These networks are inspired by the structure and function of the human brain, which consists of layers of interconnected nodes that process information. Deep learning models excel at tasks such as image and speech recognition, natural language processing, and robotics because they can learn to identify complex patterns in data.

One of the primary distinctions between ML and DL is that while ML models require manual feature engineering, DL models are capable of automatically learning hierarchical representations of data. This means that deep learning models can recognize intricate patterns and correlations in data without human intervention, making them more powerful and scalable.

The development of artificial intelligence can trace its roots back to Alan Turing, one of the founders of modern computing. Turing was part of a group of visionary scientists who laid the groundwork for artificial intelligence in the early twentieth century. His invention of the Turing machine, an abstract model of computation, is fundamental to modern computer science [3]. During World War II, he also developed a machine called the Bombe, which was used to decrypt enemy messages by systematically narrowing down the possible solutions based on a series of message transcriptions. This approach is somewhat akin to machine learning, which involves making model assumptions based on data (Fig. 2.1).

In a public lecture in 1947, Turing introduced the concept of a machine capable of learning from experience through his model of the "imitation game," now known as the Turing test. In this test, Turing demonstrated that it is impossible to distinguish between a machine and a human based solely on responses from the machine that had already been "learned," presenting the machine as capable of reasoning. Arthur L. Samuel later coined the term "machine learning" after developing the first machine that could learn to play checkers. During the training phase of machine learning, a data set is analyzed, a model is built, and the model's performance is evaluated using a function known as the goal or cost function. Later, John McCarthy described "artificial intelligence" as "the science and engineering of making intelligent machines" (Fig. 2.2).

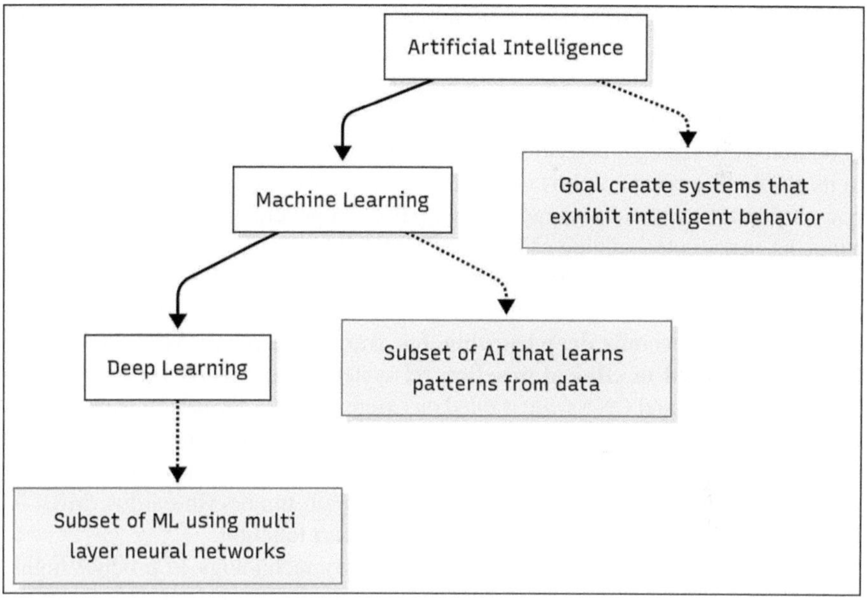


**Fig. 2.1** Schematic relationship between Artificial Intelligence, Machine Learning and Deep Learning

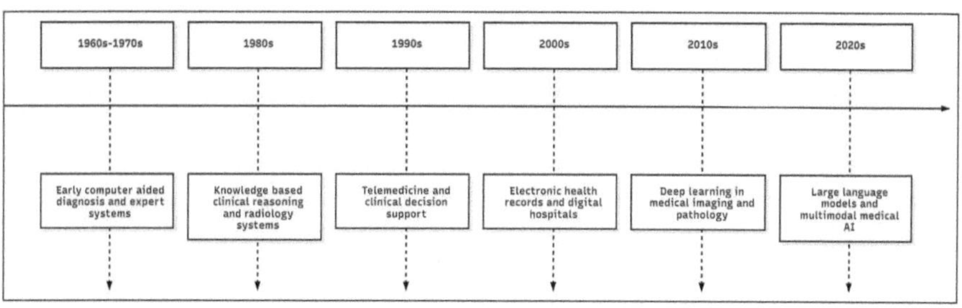


**Fig. 2.2** Timeline of AI in medicine

### *2.1.1 Overview of Medical Imaging Modalities*

**X-rays** are a form of electromagnetic radiation that produce images in varying shades of black and white. This variation occurs because different tissues in the body absorb radiation at different rates. For instance, bones, which contain calcium, absorb X-rays the most and appear white on the images, while softer tissues like fat absorb less radiation, resulting in a gray appearance. X-rays are commonly used to diagnose fractures, conduct mammograms for breast cancer screening, and perform chest X-rays to detect pneumonia. Due to their speed, low cost, and ability to identify a range of conditions, **X-ray imaging is one of the most widely used diagnostic techniques in medicine.** Large datasets of annotated X-ray images are utilized to train AI models, particularly convolutional neural networks (CNNs), which can automatically detect anomalies such as lung diseases, infections, fractures, and tumors [4]. These AI tools assist radiologists by accelerating image processing, reducing diagnostic errors, and highlighting areas of concern. Additionally, AI-powered solutions can track disease progression over time, assess severity, and triage urgent cases.

**Magnetic Resonance Imaging (MRI)** is a crucial non-invasive diagnostic tool in neurology, oncology, musculoskeletal imaging, and cardiovascular imaging due to its ability to produce high-resolution images of soft tissues. MRI scanners use a powerful magnetic field and radio wave pulses to generate body images without exposing patients to ionizing radiation. The detailed images provided by MRI are vital for diagnosis and treatment planning. However, interpreting MRI scans can be time-consuming due to the volume and complexity of the data involved. **The integration of AI, especially deep learning, has transformed how MRI data is analyzed and utilized in clinical practice.** AI systems, particularly recurrent neural networks (RNNs) and CNNs, are trained on extensive MRI datasets to perform various tasks such as quantification, image reconstruction, classification, and segmentation [5]. AI models have shown promise in brain imaging by identifying early signs of multiple sclerosis, Alzheimer's disease, and brain tumors. In cardiac MRI, AI helps detect structural abnormalities and assess heart function.

**Computed Tomography (CT) scans** use X-ray technology to produce highly detailed images of the body. The scans generate cross-sectional images of soft tissues, blood vessels, and bones, revealing **more information than standard X-rays.** CT scans are used for disease diagnosis and treatment planning, including medicinal, surgical, or radiation therapies. However, interpreting CT images can be challenging and often time-sensitive, especially in emergencies. The integration of artificial intelligence into CT imaging has significantly improved diagnostic processes, accuracy, and efficiency. AI technologies in lung imaging excel at detecting and measuring interstitial lung disorders. In neuroimaging, AI can quickly and reliably identify acute ischemic strokes and cerebral hemorrhages, which is critical in urgent situations. AI is also used in oncology to predict therapy response and track tumor growth over time. By automating repetitive tasks, AI reduces diagnostic errors and inter-observer variability while enhancing radiologist productivity [6].

These tools are especially beneficial in areas with a shortage of qualified radiologists and high patient volume clinical settings. Although challenges remain concerning validation, population-level generalization, and integration into clinical workflows, AI continues to show great potential in enhancing CT imaging and patient care.

**Ultrasound** is a popular non-invasive imaging technique that is affordable, portable, and safe. It plays a vital role in several medical specialties, including emergency medicine, cardiology, and abdominal imaging. However, the interpretation of ultrasound images heavily depends on the operator and can be limited by subjectivity and image quality. Diagnostic ultrasonography uses sound waves emitted by probes, or transducers, which operate at frequencies higher than 20 kHz, with most current transducers working at significantly higher frequencies. Typically, ultrasound probes are applied externally on the skin, but they can also be introduced internally into the gastrointestinal system, vagina, or blood vessels to enhance image quality [7]. AI applications in ultrasound imaging improve image acquisition, interpretation, and clinical decision-making by facilitating a more accurate and efficient analysis (Fig. 2.3 and Table 2.1).

### *2.1.2 Image Acquisition and Preprocessing*

**Image acquisition** is the process of capturing visual information from the physical environment and converting it into a digital image that computers can analyze. The primary goal is to transform physical scenes into digital data that can be examined, modified, or enhanced. Each image collection tool employs different methods for capturing and converting light into pixelated data. For instance, a digital camera uses a sensor to detect light and store it as digital data, while a scanner employs a different technique to capture highly accurate images of objects.

Due to its versatility and wide range of applications, image acquisition is a fundamental aspect in fields like remote sensing and healthcare. It enables advancements in AI, machine learning, and other areas. Before any processing, such as applying filters, recognizing objects, or improving image quality, can occur, the image must first be acquired. The quality of this raw data significantly impacts the efficiency of subsequent processing algorithms. Additionally, factors such as lighting, resolution, and the angle at which the image is captured can affect the accuracy of the results. The image is recorded by combining input electrical power with sensor materials that respond to specific types of energy, converting incoming energy into a voltage. The sensors' responses generate an output voltage waveform, which is then digitized to provide a digital quantity.

To obtain comprehensive data, several considerations must be made. Different sensors perform better in specific environments when capturing images. For example, some sensors offer higher color accuracy and low-light performance, while others are fast and versatile. Additionally, regular calibration of cameras, scanners, and other equipment is necessary to account for lens distortion, color errors, or

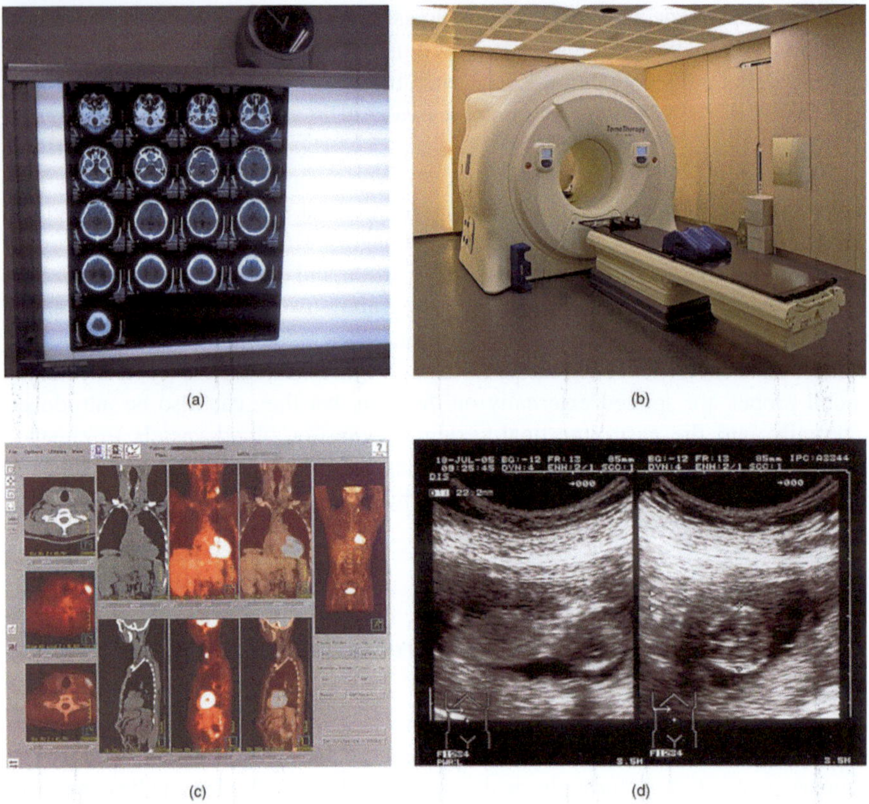

**Fig. 2.3** Medical imaging techniques (https://commons.wikimedia.org/wiki/File:113abcd_Medical_Imaging_Techniques.jpg)

misalignments, ensuring reliable and high-quality input. Using polarized light can help reduce glare in highly reflective situations, while diffused lighting can decrease shadows and reflections. High Dynamic Range (HDR) image collection methods enhance the quality of images with high contrast by capturing data in both bright and dark areas [8]. It is also beneficial to apply basic preprocessing techniques, such as gamma correction or noise reduction, during the acquisition phase to improve the raw data and lessen the burden of subsequent preprocessing (Fig. 2.4).

**Medical image preprocessing** involves a series of procedures applied to medical images to enhance their quality, standardize their format, and prepare them for further analysis. This process is crucial for analyzing and interpreting medical imaging data because it improves image quality, reduces artifacts, and prepares data for more complex analytical methods. Preprocessing medical images is essential for several reasons:

- Better image quality: It helps to reduce noise and artifacts introduced during image capture.

**Table 2.1** Comparison of imaging modalities in AI integration

| Imaging Modality | Typical Data Type | Common AI Tasks | Key Advantages | Main Challenges in AI Integration |
|---|---|---|---|---|
| X-ray/ radiography | 2D grayscale images | Automated detection of fractures, lung disease screening (e.g., pneumonia, COVID-19) | Widely available, low cost, large public datasets for training | Lower soft-tissue contrast, variable image quality across machines |
| Computed tomography (CT) | 3D volumetric images | Tumor segmentation, organ delineation, lesion detection | High spatial resolution, good anatomical detail | High radiation dose, large data size requiring heavy computation |
| Magnetic resonance imaging (MRI) | 2D/3D multi-contrast images | Brain tumor classification, multiple sclerosis lesion detection, cardiac function analysis | Excellent soft-tissue contrast, multiple sequences provide rich information | Longer acquisition time, variability across scanners and protocols |
| Ultrasound | Real-time 2D/3D images | Fetal anomaly detection, cardiac ejection fraction estimation, point-of-care diagnostics | Portable, real-time imaging, no ionizing radiation | Operator dependence, speckle noise, variable quality |
| Positron emission tomography (PET) | 3D functional images | Lesion detection, tumor staging, treatment response prediction | Functional metabolic information complementary to anatomical scans | Low spatial resolution, high cost, need for radiotracers |
| Digital pathology/ histopathology | Whole-slide gigapixel images | Cancer grading, cell detection, subtyping of tumors | Extremely detailed cellular information, high diagnostic accuracy | Very large file sizes, need for standardized staining and scanning |

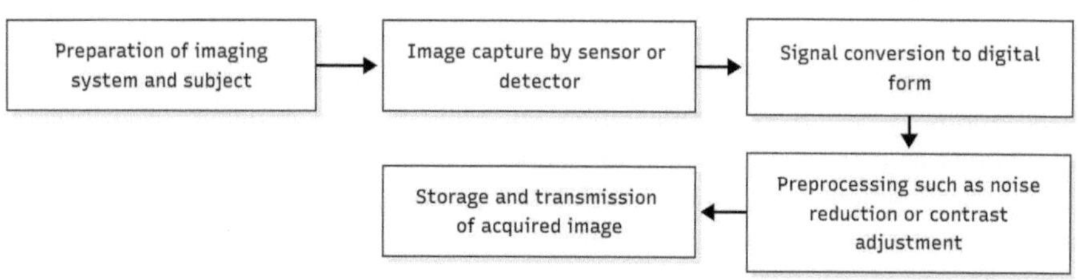


**Fig. 2.4** Main steps in image acquisition

- Standardization: It ensures consistency among images taken from different patients, scanners, or at various times.
- Enhanced analysis: It prepares images for machine learning, segmentation, and registration.

- Increased diagnostic accuracy: Improved image quality leads to more accurate diagnoses and treatment planning.

The first step in focusing on the anatomical structures of interest is **to remove the background or irrelevant areas of an image** (e.g., empty spaces or scanning labels). This reduces computational load, eliminates misleading pixels, and allows for concentration on clinically relevant portions of the image. For example, background removal separates soft tissues and skull structures from ambient noise in brain MRIs.

Medical imaging often contains noise due to weak signals, technological limitations, or environmental factors. **Noise reduction filtering** helps to clean images by removing irregular changes in intensity, which can obscure microscopic abnormalities or confuse AI systems trained to identify intricate features, such as microcalcifications in mammography. Normalization adjusts image intensities to a consistent scale, as different imaging instruments or techniques may produce varying intensity ranges. This facilitates the comparison of images from different sources and ensures consistent performance in machine learning models. For example, normalization allows AI to reliably detect tissues, such as soft tissue or bone, in CT images.

Medical images may have different spatial dimensions and resolutions depending on the modality or scanner type. **Resizing** ensures that all images are uniform in size, which is crucial for deep learning models that require fixed input dimensions. Resampling maintains the spatial relationships between anatomical components, and increasing contrast can enhance features that might otherwise be difficult to detect, particularly in soft tissue imaging [9]. Whether for clinical diagnosis or the training of AI models, each preprocessing step is crucial for enhancing picture quality, consistency, and interpretability. When combined, these methods create a vital pipeline that transforms unprocessed medical data into a strong basis for advanced diagnostic tools, ultimately improving precision, effectiveness, and patient outcomes.

**AI-based segmentation and classification** have become essential for enhancing the accuracy and efficiency of medical image analysis. Segmentation involves delineating anatomical structures or pathological regions within an image, while classification assigns labels to these regions based on learned patterns. Deep learning models, particularly convolutional neural networks (CNNs), have shown exceptional performance in these tasks. For example, the U-Net architecture is widely used for medical image segmentation, enabling precise boundary detection in modalities such as MRI, CT, and ultrasound.

In classification tasks, models like ResNet and DenseNet have been effectively applied to detect diseases such as diabetic retinopathy, pneumonia, and skin cancer, achieving diagnostic performance comparable to that of radiologists. These AI systems not only reduce inter-observer variability but also assist clinicians in early diagnosis and treatment planning, thereby improving patient outcomes. However, challenges remain regarding generalizability across diverse populations and imaging devices, emphasizing the need for robust validation and regulatory oversight.

### 2.1.3 Deep Learning in Radiology and Pathology

Deep learning, a subfield of machine learning that utilizes artificial neural networks with multiple layers, has become a transformative force in medical imaging, particularly in radiology and pathology. Its strength lies in its ability to learn hierarchical feature representations directly from raw data, eliminating the need for manual feature engineering [10]. This capability is especially important in medicine, where imaging data is extensive, heterogeneous, and often characterized by subtle visual differences.

In the field of **radiology,** deep learning models, especially **Convolutional Neural Networks (CNNs),** have demonstrated remarkable performance in image classification, object detection, and segmentation. CNNs emulate the human visual cortex and excel at recognizing spatial hierarchies within image data. For instance, the early layers of a CNN can detect edges and textures, while the deeper layers learn to recognize more complex structures, such as organs or pathological patterns.

One notable application is the **automated detection of lung nodules in chest CT scans** using 3D CNNs. These networks analyze volumetric data across slices to identify small nodules that might be overlooked by the human eye. Similarly, deep learning models have been employed in brain MRI analysis for detecting gliomas, predicting tumor grades, and segmenting areas of edema and necrosis. These models often utilize advanced architectures like **U-Net and ResNet**, which incorporate skip connections to preserve spatial information and maintain performance in deeper networks. Another emerging concept is multi-modal learning, where models learn from various imaging sources or integrate imaging with clinical data (for example, combining MRI with patient demographics or lab results). This approach can enhance diagnostic accuracy and facilitate personalized treatment planning.

**Pathology** presents its own challenges, as gigapixel images from digitized whole-slide images (WSIs) can exceed 100,000 × 100,000 pixels. Processing such massive images necessitates a technique known as patch-based learning, where the WSI is divided into smaller patches (e.g., 256 × 256 pixels) that are analyzed individually. The results from these patches are then aggregated using methods such as multiple instance learning (MIL), enabling the model to make slide-level predictions without needing pixel-level annotations.

Recent deep learning models, such as **LYNA (Lymph Node Assistant),** developed by Google, have been trained to detect micrometastases in breast cancer histopathology slides. LYNA has achieved performance comparable to that of expert pathologists, particularly in high-throughput clinical settings where diagnostic fatigue can affect accuracy. Additionally, Vision Transformers (ViTs) are a newer architecture gaining traction in pathology. Unlike CNNs, which utilize convolutional filters, transformers process image patches as sequences—similar to how they process text [11]. This allows for better modeling of long-range dependencies, which is valuable for understanding spatial relationships in tissue architecture.

Several emerging techniques are further enhancing deep learning capabilities in these domains:

- Self-supervised learning: Instead of relying on large annotated datasets, models learn from unlabeled data by solving pretext tasks (e.g., predicting image rotations or missing parts). This is particularly useful in medical imaging, where creating annotations requires considerable time and expertise.
- Federated learning: This decentralized approach enables models to be trained across multiple institutions without the need to share patient data, addressing privacy concerns while improving model generalization.
- Explainable AI (XAI): Interpretability tools such as Grad-CAM (Gradient-weighted Class Activation Mapping) highlight regions of an image that contribute most to a prediction, helping clinicians trust and understand AI-driven decisions.

Despite its potential, deep learning in radiology and pathology faces several challenges, including limited labeled data, variability in imaging protocols, and issues with generalizability across patient populations and imaging devices. Regulatory approval and clinical validation are crucial since errors in these fields can have serious consequences. Additionally, integrating these systems into existing hospital information systems (such as PACS and LIS) requires robust engineering and user-friendly interfaces. Nevertheless, ongoing advancements in model architecture, data sharing, and explainability are paving the way for the safer and more effective deployment of AI in clinical practice. As these tools evolve, they are likely to become integral components of radiologists' and pathologists' workflows, enhancing diagnostic precision and facilitating earlier, more personalized interventions.

### 2.1.4 Evaluation and Validation in Clinical Settings

The **evaluation and validation** of medical imaging technologies, particularly those involving artificial intelligence (AI) and deep learning models, are crucial steps before their deployment in clinical practice. These processes ensure that the models not only perform well on training datasets but also generalize accurately to real-world scenarios across diverse patient populations and imaging environments [12]. Without rigorous evaluation and clinical validation, even highly accurate models risk failing in practical applications, potentially leading to misdiagnosis, delayed treatments, or loss of clinician trust.

The evaluation process typically unfolds in multiple stages:

- Internal validation: This phase involves testing the model on a held-out portion of the dataset collected from the same distribution as the training data. Techniques such as k-fold cross-validation or stratified sampling are used to estimate performance metrics like accuracy, precision, recall, F1-score, AUC-ROC, sensitivity, and specificity.
- External validation: In this stage, the model is tested on entirely new data acquired from different institutions, scanners, patient populations, or imaging protocols. This is a critical test of the model's generalizability and robustness.

- Prospective validation: Unlike retrospective studies, prospective validation involves real-time testing of the model in clinical settings, where it interacts with current patient data, workflows, and healthcare providers. This phase provides insights into model reliability, integration challenges, and user experience.

**Clinical Validation in Practice**

For an AI system to be trusted in clinical settings, validation must go beyond performance metrics and encompass usability, interpretability, and clinical relevance. Clinical validation focuses on the following aspects [13]:

- Comparison with expert performance: AI systems are often compared with board-certified radiologists or pathologists. The goal is to match or exceed human-level diagnostic accuracy and consistency. Studies often utilize a "reader study" format, where experts independently analyze cases with and without AI assistance.
- Real-world impact assessment: Validation studies may assess how the tool affects actual clinical decisions. For example, does the AI model reduce diagnostic turnaround time, lower the rate of false negatives, or change treatment plans? These factors are especially important for diseases where early diagnosis significantly impacts outcomes, such as cancer or stroke.
- Workflow integration and human-AI interaction: An accurate AI model will not be adopted if it is not intuitive to use or disrupts clinical workflow. Clinical validation also evaluates how well the model integrates with existing hospital systems like PACS (Picture Archiving and Communication Systems), EHRs (Electronic Health Records), and LIS (Laboratory Information Systems).

During evaluation and validation, numerous metrics and methods are typically employed:

- Confusion matrix derivatives: These include sensitivity (true positive rate), specificity (true negative rate), and precision (positive predictive value), which are especially crucial in clinical settings where false positives and false negatives can have serious consequences.
- Receiver operating characteristic (ROC) Curves and AUC: ROC curves evaluate the trade-off between sensitivity and specificity across various thresholds. The Area Under the Curve (AUC) serves as a common metric to compare diagnostic performance.
- Dice coefficient and Jaccard Index: Primarily used in segmentation tasks, these metrics quantify the overlap between predicted and ground-truth regions of interest.
- Kaplan-Meier curves and hazard ratios: In predictive modeling for patient prognosis, such as survival analysis, these tools facilitate the evaluation of time-to-event predictions.
- Bland-Altman plots: These are utilized to assess agreement between the AI system and human experts or between different modalities (e.g., comparing AI-estimated tumor volume with manual measurements).

Before being deployed in clinical settings, AI tools must undergo scrutiny from regulatory agencies. In the U.S., the Food and Drug Administration (FDA) has developed guidelines for the evaluation of AI/ML-based Software as a Medical Device (SaMD). Validation studies submitted to the FDA must include:

- Clear evidence of clinical utility.
- Proof of model performance across varied demographics and settings.
- Demonstration of cybersecurity, data privacy, and patient safety measures.
- Robust documentation of the software development life cycle and change control processes, especially for continuously learning systems.

In the European Union, the Medical Device Regulation (MDR) applies similar standards for AI-powered diagnostic tools.

### *2.1.5 Integration with PACS and Healthcare Systems*

**Picture Archiving and Communication Systems (PACS)** play a vital role in modern medical imaging by facilitating efficient storage, retrieval, and sharing of images. By replacing physical films with digital formats, PACS significantly reduce delays in diagnostics and foster collaboration across different medical disciplines [14]. The integration of PACS with healthcare systems ensures that imaging data is accessible across various platforms, enhancing workflow and continuity of care.

Effective integration relies on standardized protocols for communication of images and data:

- **DICOM** (Digital Imaging and Communications in Medicine): This standard ensures consistent formatting, storage, and transmission of imaging data, enabling various imaging modalities and systems to communicate effectively.
- **HL7** (Health Level Seven): This standard provides a framework for exchanging clinical and administrative data among healthcare applications, ensuring structured and standardized messaging.

Combining DICOM for imaging and HL7 for textual data promotes seamless interoperability between PACS, Radiology Information Systems (RIS), and Hospital Information Systems (HIS).

Integrating PACS with **Electronic Health Records (EHRs)** allows clinicians to access imaging studies alongside lab reports, clinical notes, and other patient information within a unified interface. This integration supports timely decision-making and reduces redundancy in care delivery. For example, cloud-based PACS can synchronize imaging data with cloud-hosted EHRs, facilitating rapid remote consultations and collaborative diagnostics.

Cloud PACS has gained popularity due to its **scalability, accessibility, and cost-effectiveness**. It provides secure access to imaging data from any location, which is essential for remote diagnoses and obtaining second opinions, particularly in underserved or rural areas. This capability is particularly valuable in telemedicine, where speed and accessibility are crucial for delivering quality care (Fig. 2.5).

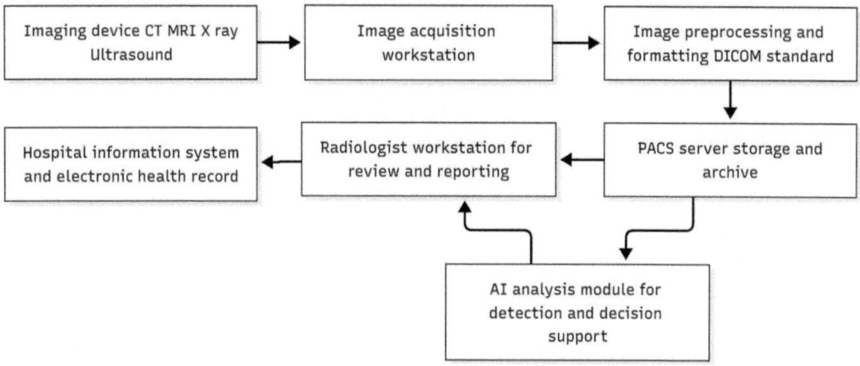


**Fig. 2.5** Integration with PACS

The **integration of Artificial Intelligence (AI) into PACS** is increasingly common, as it supports image analysis, prioritizes critical cases, and assists in diagnostic interpretation. AI tools can pre-screen studies for abnormalities, reduce false negatives, and highlight urgent findings, thereby enhancing workflow efficiency and improving patient outcomes.

Incorporating PACS into healthcare infrastructure requires robust security and privacy measures. Compliance with regulations such as **HIPAA (USA)** and **GDPR (Europe)** is essential. Common safeguards to protect sensitive medical data include encryption, role-based access control, and detailed audit logs.

Despite the advantages, integration presents several challenges:

- Standardization: Older systems may not fully support modern interoperability standards, necessitating upgrades or middleware solutions.
- Data migration and compatibility: Transferring data between platforms, especially during system updates, requires careful planning.
- Cost and resource allocation: Initial investments, staff training, and ongoing maintenance can place significant strain on institutional budgets.

Future trends are anticipated to include broader AI adoption, real-time decision support tools, more universal standardization (such as FHIR), and the expansion of mobile and cross-platform access to imaging tools. As artificial intelligence (AI) continues to transform medical imaging and diagnostics, its future development will be shaped not only by technological advancements but also by the ethical, legal, and societal responsibilities that accompany these innovations. The intersection of AI and human health necessitates careful foresight, critical evaluation, and proactive governance to ensure these systems enhance care while minimizing new risks or inequalities [15]. A promising avenue for AI in medical imaging is **multimodal integration,** the ability to combine various data sources, such as radiologic scans, histopathology slides, genomics, laboratory results, and clinical records, into a unified diagnostic tool. Multimodal AI can provide a more comprehensive understanding of a patient's condition, leading to improved disease stratification, earlier diagnosis, and personalized treatment plans. For instance, integrating PET-CT scans with genetic mutation data can help oncologists predict tumor behavior more accurately and select targeted therapies (Fig. 2.6).

Inspired by advancements in natural language processing (NLP), the field of medical imaging is witnessing the emergence of foundation models, large-scale deep learning architectures trained on extensive datasets that can be fine-tuned for various tasks. Models like **BioGPT** and **MedCLIP** are designed to interpret both images and clinical language, paving the way for universal diagnostic assistants. These systems are expected to outperform task-specific models by adapting more quickly to new imaging tasks, even with limited labeled data, due to transfer learning (Fig. 2.7).

AI is increasingly evolving from a post-analysis tool to a real-time clinical assistant. In the near future, radiologists may collaborate with AI systems that provide immediate feedback during procedures such as interventional radiology, ultrasound scanning, and intraoperative imaging. For example, AI can instantly detect bleeding or tumor margins during surgery, offering guidance in critical moments. This "**augmented intelligence**" approach enhances the clinician's decision-making process without compromising human agency.

Data privacy remains a significant concern, especially given the sensitivity of patient health records. Traditional centralized AI training often requires transferring medical data to a central server, increasing the risk of data breaches. A promising solution is federated learning, which allows AI models to be trained locally on hospital servers without moving the actual data. Only model updates are shared across sites, enabling collaboration among institutions while adhering to strict data protection regulations such as GDPR and HIPAA. This decentralized approach also

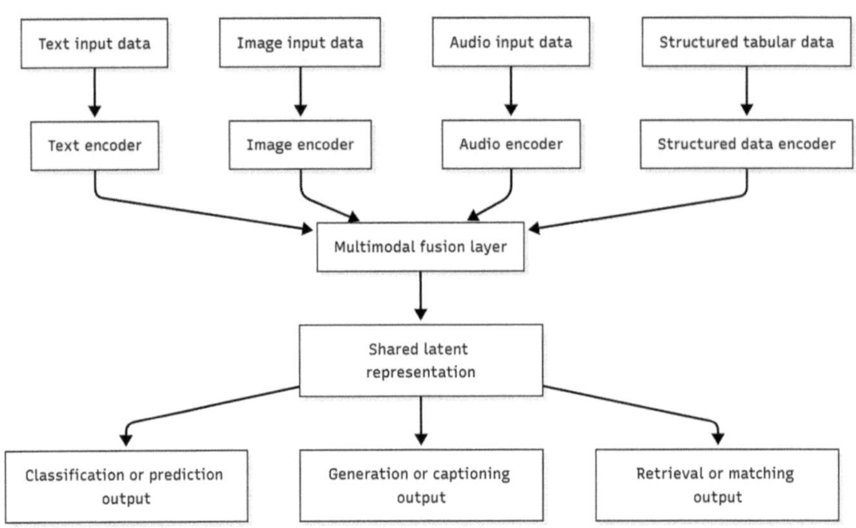


**Fig. 2.6** Schematic of a multimodal AI model

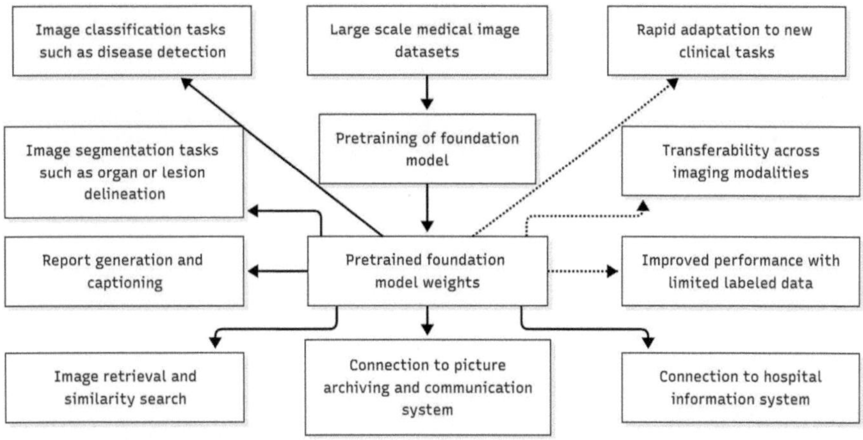


**Fig. 2.7** Overview of foundation models in medical imaging

enables the inclusion of rare diseases from smaller centers that might otherwise be excluded due to data-sharing restrictions.

A critical concern in AI development is **algorithmic bias,** which occurs when models produce unequal outcomes due to unbalanced training data. For example, skin lesion detection systems trained predominantly on lighter skin tones may perform poorly on patients with darker skin [16]. A 2018 commentary by Adamson and Smith highlighted that AI could unintentionally exacerbate healthcare disparities if inclusivity is not considered in the data pipeline from the start. Ensuring diversity in datasets is crucial for developing equitable diagnostic tools (Fig. 2.8).

Many AI systems function as "black boxes," producing outputs without clear reasoning. In critical domains such as oncology or neurology, clinicians must understand how an AI system reached its conclusions. Techniques like Grad-CAM, saliency maps, and attention visualization are being used to provide interpretability. These tools enhance trust and adoption among clinicians, enabling AI to serve as a second opinion rather than a mysterious oracle. Explainability is not only a technical necessity but also an ethical requirement for patient safety.

The **General Data Protection Regulation (GDPR)** in the European Union and the **Health Insurance Portability and Accountability Act (HIPAA)** in the United States govern the collection, use, and sharing of health data. AI systems must adhere to these regulations, ensuring:

- Data minimization: Only the necessary data is processed.
- Anonymization or pseudonymization: Data should be stripped of personally identifiable information.
- Right to explanation: Under Article 22 of the GDPR, patients have the right to an explanation regarding automated decisions that affect them.

Failure to comply with these regulations can lead to severe legal consequences, data misuse, or a loss of public trust. In 2021, the World Health Organization (WHO)

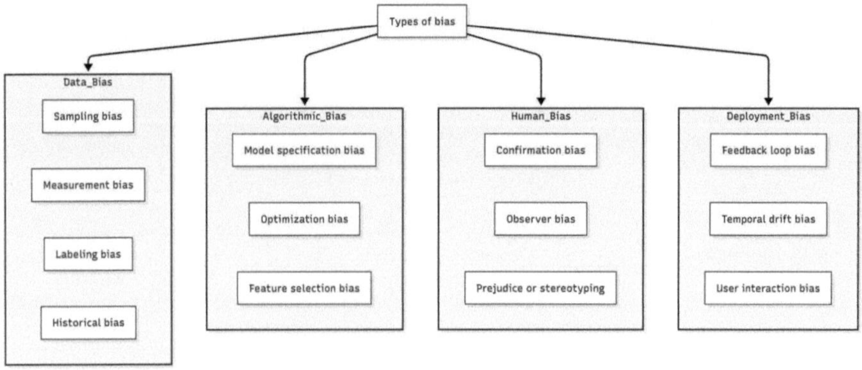


**Fig. 2.8** Types of bias

released a comprehensive guidance document outlining six key principles for ethical AI in healthcare:

1. Protect Autonomy
2. Promote Human Well-Being and Safety
3. Ensure Transparency
4. Foster Accountability
5. Ensure Inclusiveness
6. Promote Responsive and Sustainable AI

## 2.2 Intelligent Systems in Medical Device Performance Assessment

**Medical devices (MDs)** prove to be the backbone of the healthcare system nowadays. Given their high significance in maintaining human health, the processes of manufacturing, marketing and usage have been regulated globally and nationally. According to the definitions and requirements stated in directives and regulations, as well as international standards, activities related to MDs can be divided into **pre-market processes** and **post-market surveillance** (PMS).

For instance, in the European Union (EU), pre-market and PMS processes have been defined by **Medical Device Directives (MDD)** since 1992 and recently updated by **Medical Device Regulation (MDR).** The directives and regulations define all aspects from ideation, design and development phase, up to testing, approval and certification before the production process, production itself and PMS of medical devices. To support the implementation of directives and regulations international standards have been developed. For instance, according to the ISO 13485 standard, which manufacturers are obliged to comply with, during the pre-market phase, the manufacturer who produces MDs needs to demonstrate its ability to provide MDs and related services that consistently meet customer and applicable regulatory requirements. During the development phase, MD functionality, safety and performance needs to be designed to satisfy multiple requirements defined in a variety of international standards such as basic safety standards, electro compatibility standards, biocompatibility standards, environmental protection standards, etc. Certification process confirms conformity of the producer and the device itself to these requirements [17]. In the EU **CE mark** is introduced and serves as an administrative marking that indicates conformity with health, safety, and environmental protection standards for products sold within the **European Economic Area (EEA).** It is issued by the **European Notified Bodies**, organizations appointed by EU countries to assess the compliance of products before they are placed on the market (Fig. 2.9).

Generally, in the countries worldwide, MDs undergo a distinctive process from ideation, design and development phase, up to testing, approval and certification before the production, the production itself and PMS of MDs. Administrative marking that indicates conformity with health, safety, and environmental protection standards is introduced differently from country to country. For example, in China, CCC (China Compulsory Certification) marking is used for labeling of approved devices. Japan uses PSE (Product Safety Electrical Appliance & Material) marking. In Australia, there are various types of CE markings, or their equivalents used. Since leaving the EU, the United Kingdom (UK) has introduced the new mark UKCA (United Kingdom Conformity Assessed) which is enforced from January 1st, 2021. Canada enforces its own CSA (Canadian Standards Association) marking [18]. In all regions, the manufacturer governs the overall process from ideation to market, whereas independent authorized third parties check for compliance, give approval for commercialization and certification before placing the device on the market. So,

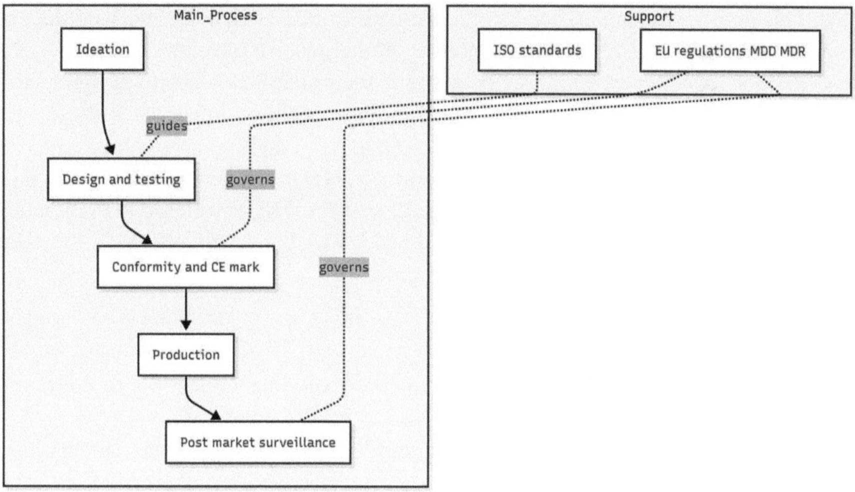


**Fig. 2.9** EU medical device regulatory flow

it is evident that MDs are produced in a strict process harmonized with international standards, and put on the market after authorization and marking.

After market placement a PMS mechanism is initiated. PMS is a collection of processes and activities used to monitor the performance of a medical device once it is placed in healthcare institutions. These activities are designed to generate information regarding the utilization of the device, to expediently identify device design and/or usage problems and accurately characterize the device behavior in practice. Despite the fact that premarket processes are very well defined and standardized and that PMS is conducted by manufacturers or distributors as defined within directives and regulations, it is not unlikely for MDs to cause an error during their usage resulting in patient injury or in the worst-case scenario resulting in patient death. The fact is that there is no difference in medical devices between countries however their performance differs which is noticeable by the number of incidents reported. So, if medical devices are the same the only cause of such state can be found in different approaches to management including preventive service and surveillance. This indicates that current PMS processes possibly have issues.

The **Global Harmonization Task Force (GHTF)** discussed PMS and defined conformity assessment principles for MDs with a scope to present and introduce a problem of non-overseen medical devices. They suggested procedures that could be performed by manufacturers to maintain a proper function of their devices on the market. Their system was based on establishing a **quality management system (QMS)** before the device is even placed on the market, to ensure that the manufacturer will in fact proceed with conformity assessments even after the device is marketed. This proposed system included handling of the complaints by customers, handling the device malfunction reports and providing preventive and corrective actions for the device maintenance. Conclusions were drafted in a report, which defined the specific tasks needed for PMS in the industry and discussed how the requirements for each task could be harmonized across regulatory environments [14]. This report was later updated by the **International Medical Device Regulators Forum (IMDRF)** to elaborate on reporting guidelines for adverse events. IMDRF has provided a document with a scope to guide their members when to exchange the information about the device, what procedures to follow during this phase and which forms to use to fill a medical device report.

In 2020, the **World Health Organization (WHO)** published a draft guidance of a proposed PMS plan to be done by the manufacturers. Their proposed plan includes information about how and when to do PMS and what to do based on the data obtained. Their plan consists of following points:

- Manufacturer must state which types of their medical devices are undergoing PMS;
- Manufacturer should state the exact method of data collection through the device's lifecycle;
- Manufacturer must state the exact method of data analysis on the obtained information;
- Manufacturer must use the obtained results in a way to improve its development phase and decrease possible risks associated with.

### 2.2.1 Performance Evaluation

Collection of information regarding the utilization of the device and the device behavior in practice is the third key component of a PMS mechanism. According to the current regional/national legislation this part of the PMS mechanism is a formal obligation of the manufacturer/distributor. The study by Ross et al. reports that out of 47 manufacturers required to approach performance inspection of their MDs, only 22 responded which is less than 50%. This report is a sound for alarm as it shows that manufacturers are not so consistent in MD performance evaluation once the MDs are placed in healthcare institutions [19]. The performance evaluation is done during the preventive maintenance, assuming that the healthcare institution is not doing that 'in-house' or during corrective maintenance when some damage potentially has already been done. Also, given that manufacturers have different distributors in different countries the collection of information regarding the utilization of the device and the device behavior in practice is aggravated. This study highlights an important issue in the PMS mechanism, and that is the lack of harmonized evidence-based methodology that will allow collection of MD performance data once they are used in healthcare institutions on a regular basis in order to assess the device behavior in practice and in different environments [20]. This has already been recognized by the professional community.

This is seen by the results of existing research focusing on MD performance evaluation. For instance, an analysis of results of PMS in the US, conducted in 2012, found that, most commonly, recalls on the US market were due to software failures in medical devices. In a study it has been reported that 30% of tested mechanical ventilators and infant incubators are not operating properly and should be serviced, recalibrated and/or removed from daily application. 13.84% of tested anesthesia machines and 14.91% of defibrillators device performance is not in accordance with requirements and should either have the device removed from use or scheduled for corrective maintenance. On the other hand, for dialysis machines, the results show that 12.6% of inspected devices do not meet electrical safety requirements or have performance outside the specifications [21, 22]. Specifically, 2% of tested devices did not pass the safety inspection in accordance with IEC 60601. For approximately 22.64% of devices that do not meet the performance specification, malfunction of heating systems was detected. Additionally, 11.32% of devices from this group had performance that was not in accordance with device specifications although malfunction was not reported.

These specific studies on performance evaluation of MDs which have been used and managed within healthcare institutions best show how much the devices actually deviate from the reference values and how important it is to develop harmonized evidence-based methodology that will allow collection of MD performance data once they are used in healthcare institutions on a regular basis in order to assess the device behavior in practice and in different environments and prevent adverse events.

The conducted review study resulted in following PMS gaps recognized [23]:

- Limited function of existing medical device registries as currently they have administrative purpose—only facilitating registration and import of medical devices;
- Adverse event reporting mechanism is managed differently globally with varying response of stakeholders;
- MD performance evaluation reveals significant variation in MD performance;
- Utilization of MDs and their performance evaluation currently cannot be assessed with great certainty since data collection and performance evaluation is not done in the same manner through healthcare institutions.

The PMS mechanism is therefore the whitespace for further improvements but with direct impact on quality of care in healthcare institutions and patient safety. Some countries have bridged the identified issues regarding PMS and MD performance evaluation by expanding the regulatory framework and introducing MDs within legal metrology.

### 2.2.2 *Intelligent Maintenance Management Systems for Medical Devices*

As is known, standards are suggested practices produced by subject matter experts with the intention of describing the best feasible way to achieve a specific end goal, increase efficiency, and eliminate product failures. Therefore, to undoubtedly ensure MD safety through their usage in healthcare institutions **harmonization of their performance evaluation during PMS** is needed. Harmonized **evidence-based conformity assessment of MDs during PMS** relying on traceability of medical device measurements will contribute to higher reliability of MD performance and consequently to higher reliability of diagnosis and treatments (Fig. 2.10).

In an environment where healthcare institutions struggle to balance cost-effectiveness and safety, it is evident that the traditional approach to healthcare management is becoming insufficient to address the growing needs. This issue is becoming more and more recognized worldwide and pilot project for evaluation of implementation of different solutions based on legal metrology were conducted in the past years.

Leveraging the power of collected data largely depends on the quality and harmonization of the collected data. Thus, in order to construct artificial intelligence enabled models for performance prediction of medical devices, it was of utmost importance to standardize the methodology of post-market surveillance of specific devices, as laid out in [3]. So, having a comprehensive database and following the standard practice of utilization of a variety of machine learning (ML) techniques, for the development of predictive models in biomedicine and for MD management the authors represent a research of developing classifiers (Expert systems) based on ML techniques and use them for MD management optimization in healthcare institutions.

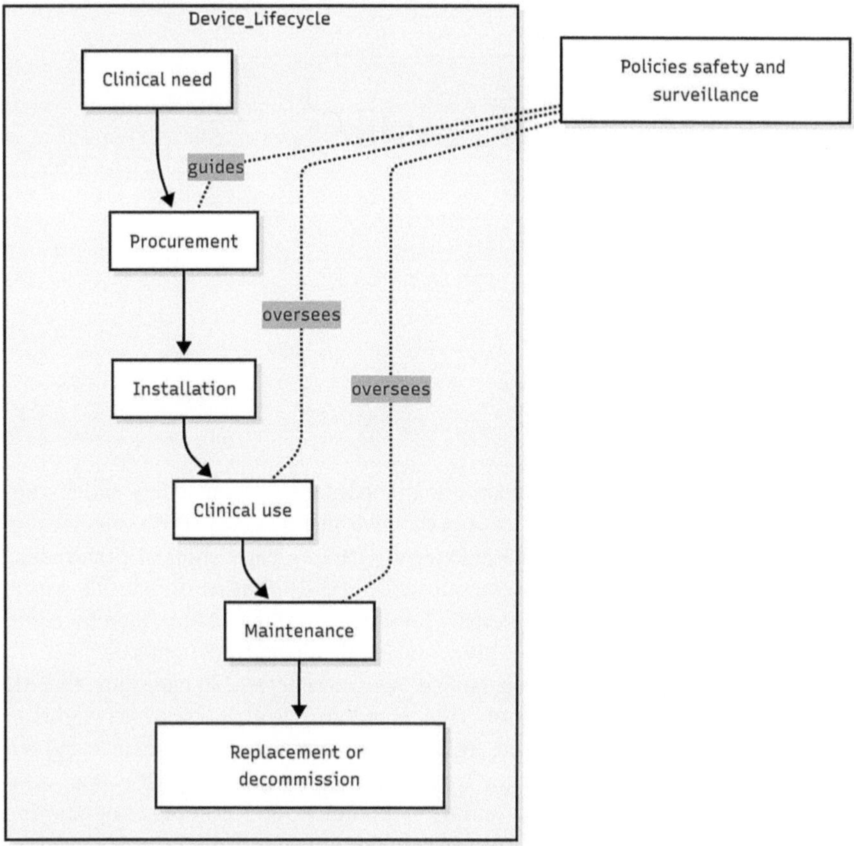


**Fig. 2.10** Usage of medical device in healthcare institutions

Advanced **intelligent medical maintenance information system (IMMIS)** platforms can even build **digital twins of maintenance tasks**, simulating real-time device data, operating conditions, and maintenance procedures. AI-driven chatbots and digital assistants are increasingly being deployed to support technicians, offering on-demand troubleshooting or step-by-step repair guidance on mobile devices. Over time, such systems may transform IMMIS from passive data recorders into proactive decision-support tools. As datasets grow and algorithms mature, this trend toward smarter maintenance is expected to intensify. For end-users, intelligent systems can also deliver clear, interactive instructions, enhancing correct device use, improving safety, and reducing operational errors (Fig. 2.11).

Recent research has increasingly explored the potential of artificial intelligence (AI) to enhance the management and maintenance of critical medical devices, particularly through the integration of machine learning (ML) approaches into **computerised maintenance management systems (CMMS)**. This shift stems from the recognition that, despite regular scheduled maintenance, a substantial number of

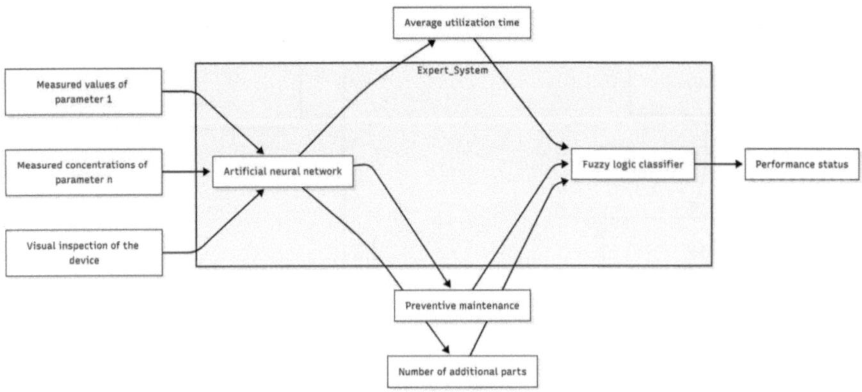


**Fig. 2.11** Expert system

equipment failures continue to occur, often jeopardising patient safety and disrupting clinical workflows. Conventional preventive maintenance (PM) strategies are largely time-based, which may not adequately capture early signs of performance degradation. AI-driven predictive maintenance (PdM) aims to bridge this gap by leveraging operational data to anticipate failures before they occur, thereby reducing downtime, improving resource allocation, and enhancing patient safety.

**Infant incubators** were among the first devices analyzed. In one study, five ML algorithms—Naïve Bayes, Decision Tree, Random Forest, K-Nearest Neighbour, and Support Vector Machine—were trained to classify device performance and predict potential failures. The Decision Tree model achieved the highest overall accuracy at 98.5%, demonstrating strong suitability for proactive monitoring. Complementing this, a hybrid expert system combining an artificial neural network and fuzzy logic classifiers achieved 95.65% accuracy, 97.23% sensitivity, and 88.89% specificity, emphasizing that complex AI systems can successfully model subtle failure patterns in devices critical for neonatal care.

**Defibrillators** were similarly examined. Using CMMS data collected from routine inspections, multiple ML algorithms were tested for their ability to classify devices as operational or at risk of failure. Decision Tree and Random Forest models delivered high accuracies (up to 99%), demonstrating the feasibility of predictive maintenance to minimize device downtime and ensure readiness in life-critical interventions. The study also illustrated that regression models could quantify error magnitude, providing a more nuanced understanding of device degradation that goes beyond binary pass/fail classification.

**Mechanical ventilators** were investigated to assess high-risk life-support equipment. Using a dataset encompassing multiple ventilator models and manufacturers, the study compared classifiers including Naïve Bayes, Decision Tree, Random Forest, KNN, SVM, and XGBoost. Ensemble models like Random Forest and XGBoost achieved predictive accuracies approaching 99%, highlighting their capacity to handle heterogeneous data and complex feature interactions. These models also allowed early identification of devices that might fail soon, supporting predictive maintenance strategies that could prevent clinical incidents.

**Infusion pumps and patient monitors** were also studied, following a similar workflow of data preprocessing, ML training, and performance evaluation. The studies consistently reported high classification accuracy, sensitivity, and specificity, confirming that predictive models can reliably distinguish between compliant and non-compliant devices. Notably, the research revealed that Decision Tree and Random Forest models often provide a favorable balance between predictive power and interpretability—a critical consideration for regulatory and clinical acceptance.

Across all device types, these studies illustrate several key insights:

1. **Quality and standardization of inspection data are essential**. ML models perform best when inspection records are accurate, complete, and standardized across devices, institutions, and manufacturers.
2. **Predictive maintenance can shift healthcare institutions from reactive to proactive strategies**, reducing device downtime and enhancing patient safety.
3. **Ensemble methods, such as Random Forest and XGBoost, consistently deliver strong performance**, but simpler models like Decision Trees remain attractive due to their interpretability.
4. **Hybrid AI systems**, combining neural networks with fuzzy logic, can capture subtle nonlinear failure patterns that may be missed by conventional algorithms.

Beyond **predictive analytics,** AI integration promises to transform maintenance management systems from reactive record-keepers into proactive, intelligent decision support platforms. Intelligent scheduling algorithms can dynamically prioritise work orders based on device criticality, fault severity, and technician availability, ensuring that urgent issues are addressed first. Advanced systems, such as IMMIS, are being designed to generate digital twins of medical device maintenance tasks—virtual models that replicate real-time operational data, environmental conditions, and maintenance activities. Such digital twins enable simulation-based planning, risk assessment, and continuous performance tracking. In addition, AI-powered conversational agents are increasingly being deployed to support maintenance personnel [15]. Embedded within CMMS interfaces, these agents can provide instant troubleshooting guidance, offer context-specific repair instructions, or even guide technicians step-by-step through repair procedures using augmented reality on mobile devices. Over time, as these systems learn from large-scale data and technician feedback, they are expected to evolve into autonomous advisers capable of optimising maintenance strategies, predicting device lifespan, and recommending procurement decisions. For end-users, intelligent systems could deliver interactive operating instructions and real-time safety prompts, reducing misuse and enhancing operational reliability.

Overall, these developments signal a paradigm shift in medical device management. As AI models become more sophisticated and as larger, high-quality datasets become available, maintenance systems are likely to transition from static, reactive frameworks to dynamic, self-optimising ecosystems. This evolution holds significant implications for improving device safety, reducing healthcare costs, and enabling more resilient clinical infrastructure.

## 2.3 Diagnostic Support Systems and Clinical Decision Making

**Clinical Decision Support Systems (CDSS)** play a crucial role in modern healthcare by assisting medical professionals in making informed decisions. They enhance patient outcomes and safety, increase productivity, and promote evidence-based practices. By integrating CDSS into clinical workflows, healthcare can become more efficient and centered around the patient.

CDSS serves as an essential tool for both healthcare providers and organizations aiming to improve clinical care. However, successfully deploying these systems can be challenging. Therefore, it is important to assess the readiness of healthcare organizations to utilize such technologies before implementation. CDSS is vital for several reasons, including enhanced patient safety, more accurate diagnoses, improved treatment plans, and more consistent care. It also assists healthcare professionals in navigating complex decisions. The objective of this study is to develop a model that helps healthcare organizations identify the challenges associated with CDSS deployment and evaluate their preparedness for such implementation in a comprehensive and multifaceted manner.

### *2.3.1 Clinical Decision Support Frameworks*

In an effort to provide high-quality healthcare services, the quality of data and the appropriate use of clinical information are crucial. Hospital decision-making relies on data collected throughout clinical procedures, which is then analyzed through the cognitive abilities of medical professionals. Clinical Decision Support Systems (CDSS) are valuable tools that enhance patient safety and reduce clinical practice errors by delivering information to clinicians that is intelligently filtered and presented at optimal times to improve health outcomes [20]. Ideally, the information provided by CDSS should reflect the real decision-making process and cognitive efforts of physicians.

Thirty years ago, in the 1980s and early 1990s, there was an open debate regarding how computers should generate diagnoses. While this discussion persists today, remarkable advancements in machine learning now offer new opportunities for targeted and accurate clinical predictions and recommendations. It is essential to consider the annotation of clinical attributes that are often used together to drive specific decisions. CDSS and their underlying algorithms should not rely on static, in vitro methods. Instead, they should make dynamic predictions, facilitate interactions with clinicians based on their feedback, and take into account the longitudinal nature of physiological properties as well as disease progression (Table 2.2).

CDSS can be broadly classified into two categories:

1. **Knowledge-Based CDSS:** These systems rely on a collection of clinical principles and rules encoded by human professionals. They are typically driven by inference engines or decision trees and function using an "if-then" logic structure. For example, a system that alerts a physician if a prescribed medication has a known interaction with another medication in the patient's records is an instance of a knowledge-based CDSS.
2. **Non-Knowledge-Based CDSS:** The main distinction of these systems is their use of machine learning models. Instead of relying on predetermined if-then rules, they discover patterns in historical data and learn from previous experiences. The two most commonly employed methods in these CDSS are Genetic Algorithms (GA), which mimic Charles Darwin's theory of natural selection by

**Table 2.2** Significance of each principle and CDSS design considerations

| Principle | Significance in Clinical Decision Support Systems (CDSS) | Key Design Considerations |
|---|---|---|
| Accuracy and reliability | Ensures recommendations are clinically valid and evidence based, fostering trust among healthcare professionals | Use high-quality validated data sources, rigorous model evaluation, continuous performance monitoring |
| Transparency and Explainability | Clinicians must understand how the system arrives at its recommendations to support informed decision making | Provide clear reasoning or interpretable outputs (e.g., feature importance, rule logic, explanation interfaces) |
| User-centered design | Improves usability and adoption by aligning with clinical workflow and minimizing cognitive burden | Conduct stakeholder interviews, iterative prototyping, integrate seamlessly with electronic health records and PACS |
| Data privacy and security | Protects patient information and complies with regulations (HIPAA, GDPR) | Strong encryption, access controls, anonymization of sensitive data, secure data transmission |
| Integration with clinical workflow | Enables timely and actionable decision support without disrupting care | Ensure compatibility with existing hospital information systems, provide alerts and recommendations at the point of care |
| Adaptability and scalability | Allows the CDSS to evolve with new medical knowledge and scale to larger patient populations | Modular architecture, automated updates of clinical guidelines, support for multiple specialties |
| Regulatory compliance | Ensures safety and legal approval for clinical use | Follow standards such as ISO 13485, IEC 62304, FDA or CE marking requirements |
| Ethical and bias mitigation | Reduces risk of discriminatory outcomes and supports fair treatment of all patient groups | Bias detection, balanced datasets, regular audits, inclusion of diverse patient populations in model development |

generating diverse random solutions and refining them until the best one is identified, and Artificial Neural Networks (ANN), which simulate human thinking. ANN consists of interconnected "neurons" (called "neurodes") linked through weighted connections that function similarly to nerve synapses, facilitating signal transmission across the network.

While non-knowledge-based CDSS promise to significantly reduce healthcare costs and alleviate the burden on medical professionals, several challenges hinder their widespread adoption. These challenges include lengthy and computationally intensive training processes, the need for large datasets to improve model accuracy, and primarily, a lack of interpretability, as these systems often cannot explain the reasoning behind their decisions. Due to these drawbacks, modern CDSS are predominantly knowledge-based.

**Advantages of a Decision Support System**

- Improved decision quality: Decision Support Systems (DSS) provide complex models and detailed information on specific topics, leading to quicker and more effective decision-making. They help managers identify the best solutions by allowing them to compare various approaches to a problem.
- Efficiency and speed: DSS significantly reduces the time and effort required for decision-making by automating data collection, analysis, and reporting processes. This allows organizations to react more swiftly to emerging opportunities and changes in their business environment.
- Increased productivity: By shortening the decision-making cycle, DSS enables managers and staff to allocate more time and energy to other critical aspects of their work, rather than solely focusing on data calculation and analysis.
- Better data management: DSS integrates data from various sources to provide decision-makers with accurate, up-to-date, and consistent information, helping to prevent errors.

**Disadvantages of a Decision Support System**

- High cost: Setting up and maintaining a Decision Support System (DSS) can be expensive, with costs that extend beyond the initial investment. This includes expenses for hardware, software, data collection, customization to meet specific needs, and ongoing support and maintenance.
- Complexity: Implementing a DSS can be a complex process and is not always straightforward. It may require specialized expertise in system design and implementation, as well as training for users.
- Data quality and integration issues: The effectiveness of a DSS is heavily dependent on the quality of the data it uses. Data integration can be challenging, and poor data quality can lead to inaccurate outputs.
- User resistance: Some users may resist adopting a DSS due to fear of change, concerns about job security, or a lack of knowledge regarding the benefits of the system. Overcoming this resistance often requires effective change management and thorough user training.

- Dependence on technology: A significant drawback of a DSS is its reliance on technology for decision-making. If the system malfunctions, or if users become overly dependent on its outputs without applying their own judgment and experience, the impact can be detrimental.
- Privacy and data security issues: Implementing a DSS often involves collecting substantial amounts of sensitive data. Individuals and organizations may feel that storing their information in a DSS compromises their privacy. Thus, it's essential to have robust security measures in place and to comply with relevant data protection laws and regulations.
- System limitations: It is important to note that a DSS may not be suitable for all types of decision-making, especially those requiring creativity, intuition, or subjective judgment. Additionally, it may struggle with real-time data processing or responding to extraordinary situations.
- Information overload risk: DSS often handles large volumes of data, and if this information is not managed or presented effectively, it can lead to confusion among users. As a result, decision-makers might choose not to make a decision at all, rather than effectively utilizing the insights offered.
- Updating and maintenance: Keeping a DSS up-to-date with the latest models, data, and technology can be challenging. Frequent updates are often necessary to stay current, which can incur additional costs.
- Potential bias: There is a risk of bias in using a DSS. If the models and algorithms are not developed and validated correctly, they may introduce biases into the decision-making process. This is particularly critical, as biased models can lead to misleading statistics and, ultimately, incorrect conclusions.

### *2.3.2 AI and Machine Learning in Diagnostic Support*

**Big data and artificial intelligence (AI)** are revolutionizing clinical decision support systems (CDSS) in healthcare by enhancing diagnostic accuracy, optimizing treatment plans, and improving patient outcomes. Traditional CDSS primarily relied on rule-based expert systems, which were limited in scalability and versatility. With the rapid expansion of wearable technology, medical imaging, electronic health records (EHRs), and genetic data, it has become increasingly difficult for physicians to manually analyze all available patient information. Therefore, implementing AI-driven CDSS, which utilize machine learning (ML), deep learning (DL), and natural language processing (NLP), is essential for providing healthcare practitioners with real-time, personalized, evidence-based recommendations. These AI-enhanced systems can identify disease trends, predict patient decline, and support clinical decision-making with unparalleled accuracy by processing vast amounts of both structured and unstructured data.

**Big data analytics** plays a crucial role in the functioning of AI-driven CDSS, facilitating the collection, storage, and analysis of extensive healthcare datasets. Unlike traditional decision support systems that rely on static knowledge bases, AI models trained on real-world health data can adapt dynamically to new medical discoveries, patient records, and treatment outcomes. Big data technologies such as cloud-based platforms, Apache Spark, and Hadoop enable CDSS to quickly generate insights based on clinical processes through real-time processing and predictive modeling [24]. By consolidating information from diverse sources, including wearable sensors, genetic sequencing, medical imaging, electronic health records, and clinical literature, AI-powered CDSS create comprehensive patient profiles. This multi-modal approach ensures that clinicians are well-informed about a patient's condition, leading to faster and more accurate decision-making.

Despite the tremendous potential of AI in CDSS, there are several challenges that must be addressed before widespread implementation can occur. These challenges include AI prediction bias, model interpretability, data privacy, and compatibility with current medical systems. It is critical to comply with regulatory frameworks such as the GDPR and the HIPAA to protect patient privacy and promote ethical AI practices. Technologies that enhance explainability, such as LIME) and SHAP should be integrated into the architecture of AI-driven CDSS to ensure that clinicians can understand, trust, and evaluate AI recommendations. Collaborating to resolve these issues will be essential for creating reliable, ethical, and effective AI-driven CDSS solutions, involving healthcare professionals, AI researchers, and policymakers.

CDSS represent a transformative technology in healthcare, facilitating better, evidence-based decisions by physicians. Historically, CDSS depended on standardized clinical guidelines and rule-based algorithms, which were insufficient for managing complex patient data and dynamic medical scenarios [25]. However, modern CDSS have evolved into sophisticated systems capable of analyzing large volumes of real-time healthcare data, uncovering hidden patterns, and providing personalized treatment recommendations, thanks to the advancements in big data and AI.

Although challenges remain in the implementation of AI-driven CDSS, the overall benefits significantly outweigh these issues. The caliber and variety of data utilized to train AI models, especially machine learning-based models, are crucial. Instead of reducing health inequities, incomplete, biased, or inadequate datasets can produce unfair or erroneous projections. Important concerns include:

- Incomplete data: Medical records may sometimes lack crucial information, resulting in erroneous inferences or forecasts. For example, AI models' ability to detect risk factors for certain populations may be impacted by missing demographic information.
- Data bias: If AI models are trained on datasets that do not adequately reflect specific patient groups (such as minority populations, rare conditions, or underrepresented geographic locations), their accuracy may suffer for those groups. This could lead to misdiagnosis or unfair treatment.
- Data quality: Inconsistencies or mistakes in data gathering, whether due to missing records, imaging artifacts, or human entry, can introduce noise that impairs AI models' performance and diminishes their reliability in clinical contexts (Fig. 2.12).

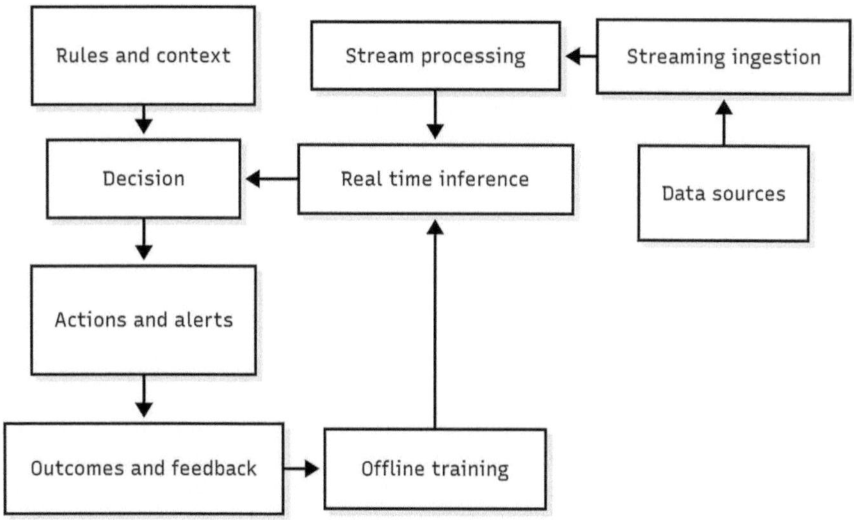


**Fig. 2.12** Real-time decision-making with AI and big data

Over the past two decades, there has been a significant increase in the use of Electronic Health Records (EHR) data in healthcare research, with many studies focusing on identifying risk factors and predicting disease outcomes. Another technique for drawing conclusions about causality is instrumental variable (IV) analysis. This approach is particularly useful when there are concerns about endogeneity, which refers to the relationship between an independent variable and the error term in a regression model. In the context of electronic health records (EHRs), instrumental variables may include factors that are not directly related to patient characteristics but influence treatment choices, such as the distance to a healthcare provider or differences in hospital treatment protocols. By utilizing these instrumental variables, IV analysis aims to identify treatment variations that are unaffected by confounding variables. While IV approaches can help address endogeneity issues, they require strong assumptions about the reliability of the instruments used, and finding dependable instruments can be challenging.

**Causal graph theory** has gained prominence in recent years as a technique for understanding the relationships between variables in observational studies. Researchers can use **directed acyclic graphs** (DAGs) to visually depict the relationships between variables, which helps identify potential biases and confounders. By constructing a DAG, researchers can more accurately estimate causal effects and determine the necessary adjustments to account for confounding. This method has been applied to visualize intricate interactions among patient characteristics, treatment variables, and outcomes using EHR data. In EHR analysis, machine learning methods are increasingly being employed for causal inference [26]. Techniques like double machine learning and causal forests offer promising alternatives to traditional statistical methods by combining high-dimensional data with complex nonlinear interactions. These methods have been shown to improve the precision of causal effect estimation, especially when dealing with large datasets containing numerous variables. Despite recent advancements, several challenges remain when applying causal inference techniques to EHR data. One significant obstacle is unmeasured confounding. Important confounders, such as socioeconomic status, lifestyle choices, and genetic predispositions, may not be fully captured in EHR systems in many healthcare settings. As a result, these hidden biases can go unnoticed, even with the most sophisticated causal inference methods. Additionally, the volume and complexity of EHR data, along with issues like misclassification, measurement error, and missing values, can complicate causal analysis.

**Real-Time Decision Making (RTDM)** refers to the process of swiftly making decisions using the information available at the moment as events unfold. It encompasses the steps taken by individuals, groups, or software when faced with urgent choices in real-time. This approach is becoming increasingly important in today's technology-driven environment, where the speed of decision-making is crucial.

Decision-makers utilize real-time data analytics to gain insights and make critical decisions rapidly. This expedited decision-making process, based on current information, enhances organizational agility and efficiency. When time is of the essence, decision-makers often rely on intuition and experience. This acquired intuition can significantly influence the decision-making process and yield positive outcomes when complemented by analytical insights. Real-time decision-making requires advanced methods and tools, such as artificial intelligence and machine learning. These technologies facilitate the rapid analysis of large datasets, help identify patterns, and enable the prediction of outcomes. In fast-paced situations, some critical decisions may be automated. However, these decisions should be genuinely autonomous and free from human intervention.

Successfully implementing real-time decision-making requires a **systematic and rigorous approach**. Key tasks include identifying which important business decisions would benefit from this method, developing a strategy, establishing key performance metrics, and regularly evaluating your decision-making processes. Other essential elements of implementation include investing in the appropriate software and technology, providing relevant training for staff, and considering potential risks along with strategies for their mitigation.

### *2.3.3 Clinical Decision Support System for Rare Disease Diagnosis*

The World Health Organization (WHO) defines a **rare disease** as an illness affecting fewer than 1.3 out of 2000 people. Currently, an estimated 400 million individuals are affected by approximately 7000 known rare diseases (RDs). Many of these illnesses are degenerative, chronic, or even fatal, leading to significant impairments in quality of life. While approximately 80% of RDs are attributed to genetic factors, other potential causes include infectious, immunological, or environmental factors [27]. Patients often endure lengthy periods—sometimes years—without a proper diagnosis due to the rarity of these diseases, which can result in delayed or missed diagnoses, particularly for individuals who present symptoms later in life. This search for an accurate diagnosis can lead to considerable suffering and limitations in patients' lives.

The geographic dispersion of RD experts complicates the situation further. Additionally, the scarcity of clinical trials and limited patient data hinder research and patient care. A promising strategy in addressing these challenges is the utilization of research networks, such as the **MIRACUM consortium** (Medical Informatics in Research and Care in University Medicine), which aggregates clinical evidence from multiple hospitals. One significant initiative arising from MIRACUM is **DISERDIS** (Diagnosis Support in Rare Diseases), a CDSS aimed at helping identify similar patients across various MIRACUM locations to facilitate diagnoses.

During the development of a CDSS, it is crucial to engage future users at every stage to ensure the system's acceptance and practical application in daily clinical settings. To achieve this, we adopted a **"User-Centered Design Process"** (UCD) for DISERDIS. Several steps were taken within the context of UCD to identify user requirements, including a scoping review of the literature on CDSS for RDs, a cross-sectional survey, qualitative studies involving expert interviews, and focus groups. Based on the findings from these studies, DISERDIS has been developed into a high-fidelity prototype that meets UCD standards but is not yet a fully functional software system [24]. The system enables users to input and view data for patients awaiting diagnosis through a web application. This data encompasses demographic information, symptoms, and family history. Furthermore, the system facilitates a patient similarity analysis by comparing data from undiagnosed patients with those previously diagnosed at different MIRACUM Rare Disease Centers (RDCs). The outcomes of this analysis are visualized in various formats, such as scatterplots, patient timeline views, and time-series charts. If similar patients are identified, users can search for experts on the specific rare disease using the "SE Atlas" platform.

## 2.4 AI-Driven Personalized Medicine and Drug Discovery

The combination of AI and biomedical science is revolutionizing the way we understand, diagnose, and treat diseases. AI is making personalized medicine— which focuses on creating treatments tailored to the unique characteristics of each patient—more effective and accessible. AI is also speeding up drug research by improving the accuracy of identifying potential chemicals [28]. By facilitating the precise identification of promising compounds, AI is playing a crucial role in accelerating drug discovery.

Two important areas where AI is significantly influencing healthcare are:

- Better categorization of patients for individualized care.
- Integration of genetic and molecular data (such as transcriptomics and genomics) to enhance our understanding of diseases and develop tailored treatments (Fig. 2.13).

Artificial intelligence is emerging as one of the most promising tools in personalized medicine and drug discovery. This includes technologies such as machine learning, deep learning, and natural language processing, all designed to analyze and interpret complex data. With these tools, AI can detect subtle yet significant patterns within vast datasets—something that would be nearly impossible to achieve manually. This capability enables doctors and researchers to better classify various disease types, predict individual risk, and develop treatments tailored to each person.

**Patient stratification,** which involves grouping patients based on shared characteristics or expected outcomes, is one of the most prominent applications of AI in healthcare. Traditionally, physicians categorized patients based on factors such as age, symptoms, or laboratory test results. However, with AI, we can analyze large volumes of data—including genetic information, electronic medical records, and imaging results—to uncover patterns that may be overlooked by human analysts. AI tools, particularly those utilizing machine learning, excel at **recognizing subtle patterns in data.** One effective method is clustering, which groups patients based on biological markers, even when they do not exhibit the same symptoms. This approach can be particularly useful for diseases like cancer, where two patients may receive the same diagnosis but respond quite differently to treatment. By identifying these distinctions early on, physicians can select medications that are more likely to be effective for each individual [29]. For instance, in cancer treatment, AI can

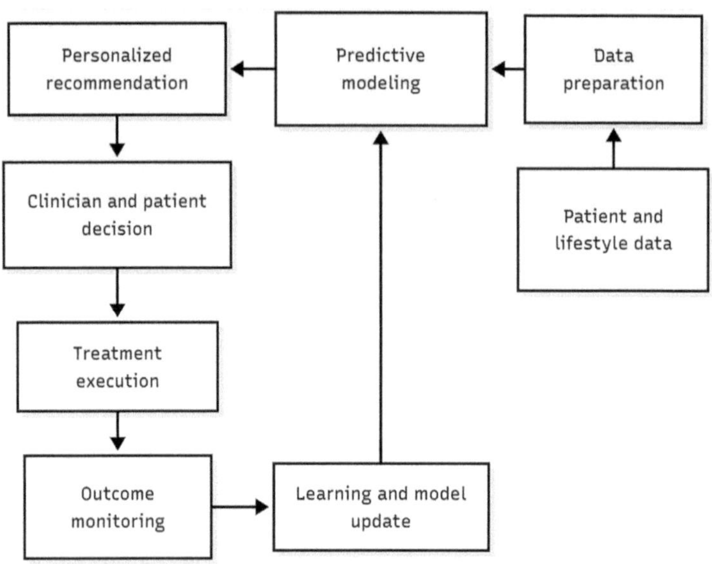


**Fig. 2.13** AI-assisted personalized treatment outcomes

predict a patient's potential response to chemotherapy by analyzing the genetic makeup of their tumor. In cardiology, machine learning is used to identify individuals at risk for heart failure, often before any symptoms appear. This represents a significant shift toward proactive, personalized care.

There are already real-world examples of this approach in action. In oncology, AI is employed to predict how a patient will respond to chemotherapy based on their tumor's genetic profile. Similarly, machine learning models are being developed in cardiology to identify individuals who may be at higher risk of heart failure before symptoms manifest (Fig. 2.14).

The primary advantage of utilizing AI in this manner is that it helps healthcare providers transition from a reactive approach, treating illness after it occurs, to a proactive strategy focused on preventing illness or detecting it early. Nonetheless, challenges remain, such as ensuring that AI does not introduce bias or make

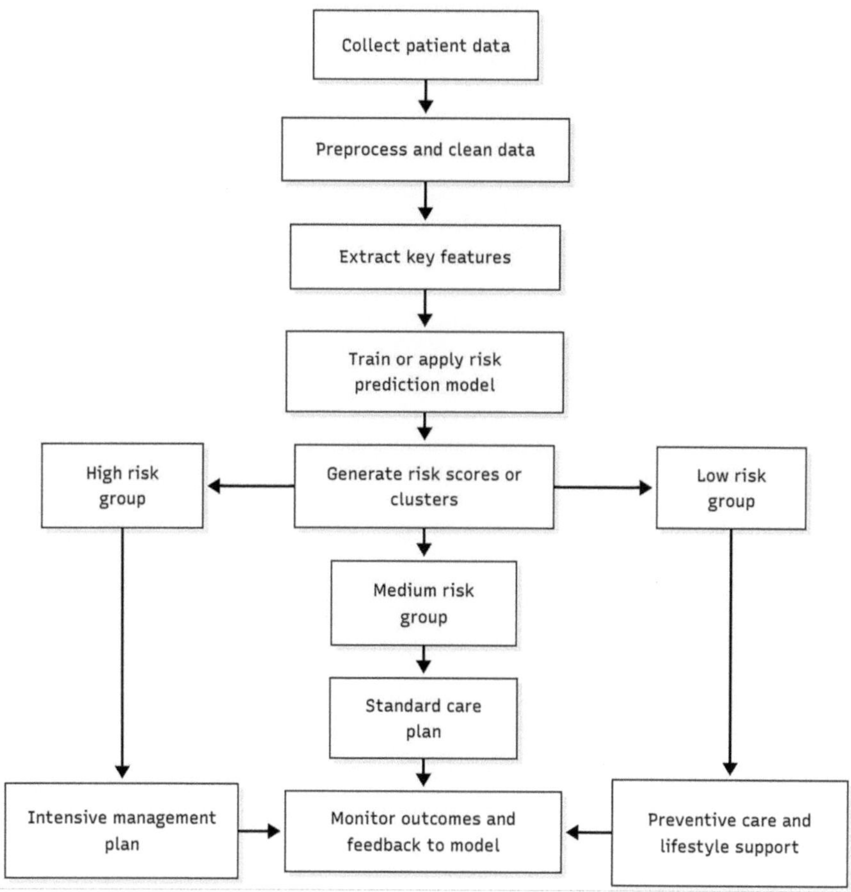


**Fig. 2.14** Patient stratification workflow

decisions based on insufficient data [30]. Despite these challenges, AI is advancing medicine towards a future where care is truly tailored to each patient.

**Genomics** is the study of an individual's DNA and all of their genes, while transcriptomics examines which genes are actively expressed in the body. When these two disciplines are combined, they provide a substantial understanding of how diseases originate, develop, and respond to treatment. This integration of various fields, including proteomics, is referred to as **multi-omics data integration** (Fig. 2.15).

For instance, AI can detect specific gene activity patterns associated with certain types of cancer or other diseases. This capability leads to a more accurate understanding of disease causation, faster diagnoses, and more personalized treatment plans. By utilizing AI to integrate transcriptomic and genomic data, researchers can identify which genes contribute to a disease and their responses under different conditions. This enhances the prediction of which treatments will be most effective for individual patients, facilitating the development of targeted therapies. However, the vast volume and complexity of this data present significant challenges. Genomic sequences can involve billions of data points, and transcriptomic profiles are often similarly extensive and more dynamic. Managing this information manually or even using traditional statistical methods is impractical. This is where artificial intelligence steps in. AI, particularly through machine learning and deep learning models, helps decipher the chaos. These systems can detect patterns and relationships in large, noisy datasets that would be difficult, if not impossible, for humans to identify independently. For example, a deep neural network can be trained to recognize which genetic variants are associated with specific diseases or how certain gene expression patterns link to a patient's likelihood of responding to particular treatments.

We can classify genomic data as static (unchanging) and transcriptomic data as dynamic (varying based on factors like splicing patterns or silencing mechanisms). Integrating this information manually would be extremely challenging, but AI models trained on both datasets can uncover valuable insights [31]. This integration is

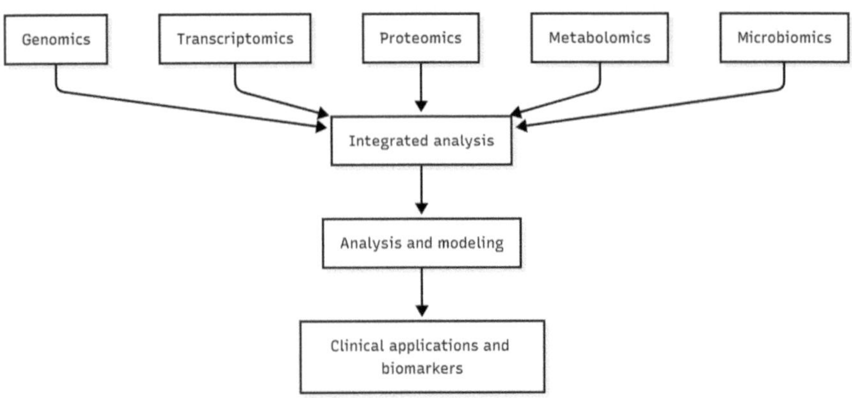


**Fig. 2.15** Multi-omics data integration in relation to one another

critically important in clinical settings and personalized medicine. Several AI algorithms are used in personalized medicine:

1. **ML algorithms** analyze large amounts of patient data to identify patterns, predict disease progression, and recommend treatments. Supervised learning models (e.g., support vector machines, random forests) and unsupervised learning techniques (e.g., clustering, dimensionality reduction) are commonly used.
2. **DL models**, such as convolutional neural networks (CNNs) and recurrent neural networks (RNNs), examine complex biomedical data, including medical imaging, genomic sequences, and electronic health records (EHRs).
3. **AI-powered big data techniques** enable large-scale processing of multi-omics data (genomics, proteomics, metabolomics) and facilitate real-time patient monitoring, thereby improving healthcare outcomes and treatment precision (Fig. 2.16).

There are two primary challenges in integrating genomics and transcriptomics data:

1. Data compatibility: Genomic and transcriptomic data often originate from different platforms, formats, and may contain various types of noise and bias. AI models must be carefully designed to address these complex issues.
2. Interpretability: Some models, particularly deep learning models, might make accurate predictions, but the reasoning behind these predictions is not always clear. Understanding these models can be challenging, especially in personalized medicine, where trust and accountability are crucial.

Large datasets, such as The Cancer Genome Atlas (TCGA), enable AI models to sift through an immense array of information, including gene activity, DNA mutations, and methylation marks on DNA, all of which influence tumor behavior. AI helps us identify distinct tumor subtypes that might otherwise appear identical. This is not

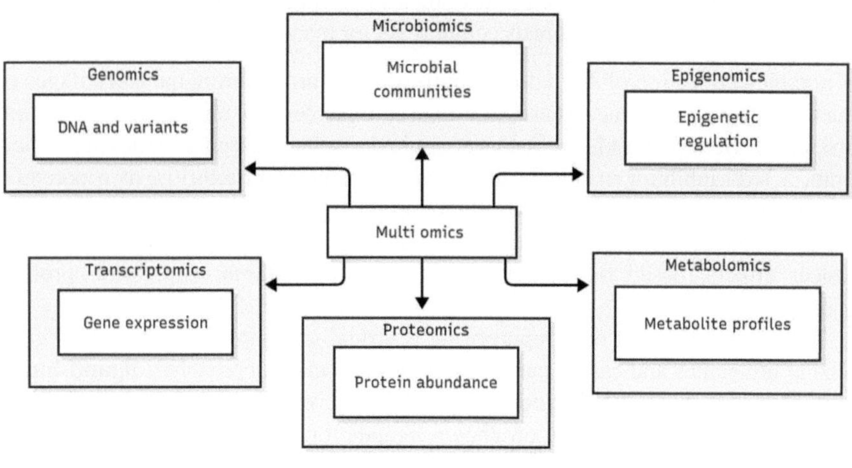


**Fig. 2.16** Multi-omics different disciplines

merely a clever technique; it enables doctors to predict the aggressiveness of cancer more accurately and select treatments tailored to each patient's specific needs.

**Drug discovery** is an inherently complex, resource-intensive, and time-consuming process. Traditional methods can take over a decade and cost billions of dollars to bring a single drug to market, with a high rate of attrition during clinical development. In response to these challenges, the biomedical research community has increasingly turned to drug repurposing and virtual screening as more efficient strategies.

**Drug repurposing** involves identifying new therapeutic uses for existing or previously failed drugs, while virtual screening utilizes computational techniques to evaluate large libraries of compounds to identify those likely to interact with a biological target. Drug repurposing, also known as drug repositioning, leverages the established safety profiles of existing drugs to discover new indications. This approach significantly shortens the timeline for clinical application, particularly during urgent health crises such as the COVID-19 pandemic. Traditional repurposing methods often relied on serendipity or manual literature curation [32]. However, artificial intelligence (AI) now enables a more systematic, scalable, and data-driven approach. Various AI methods are employed for drug repurposing:

1. **NLP algorithms** can mine biomedical literature, patents, and clinical trial records to identify potential drug-disease associations. Tools such as BERT-based models have been used to extract relational data and map interactions between drugs, genes, and diseases.
2. **Knowledge graphs** are AI constructs that navigate biomedical knowledge graphs, where entities such as drugs, proteins, and diseases serve as nodes, and their relationships are represented as edges. Graph neural networks (GNNs) have proven effective in predicting new connections, suggesting novel drug-disease links.
3. Predictive modeling using supervised learning models that are trained on known drug-target or drug-disease relationships to predict new interactions. Common models include random forests, support vector machines, and neural networks.

A notable application of AI in drug repurposing occurred during the early stages of the COVID-19 pandemic, where AI was used to screen existing drugs for antiviral properties. BenevolentAI, for instance, applied machine learning to identify baricitinib, a JAK inhibitor originally designed for rheumatoid arthritis, as a potential treatment based on its predicted ability to inhibit viral entry. Another example is the use of deep learning to repurpose metformin, a diabetes drug, for its potential anticancer effects, based on transcriptomic reversal of tumor expression profiles (Fig. 2.17).

The core method for drug repurposing is virtual screening, which involves visualizing molecules and can be categorized into two main approaches: **ligand-based virtual screening (LBVS) and structure-based virtual screening (SBVS).** In LBVS, AI models are trained on known ligands for a given target to predict new compounds with similar properties. Traditional methods rely on molecular fingerprints or descriptors. In contrast, AI models, particularly deep learning approaches

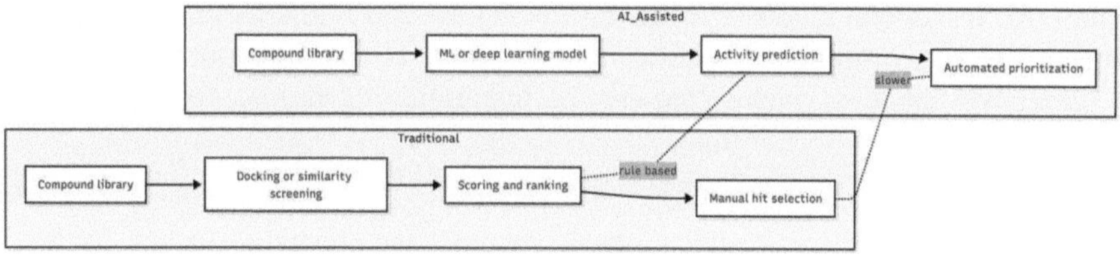


**Fig. 2.17** Traditional vs AI assisted virtual screening

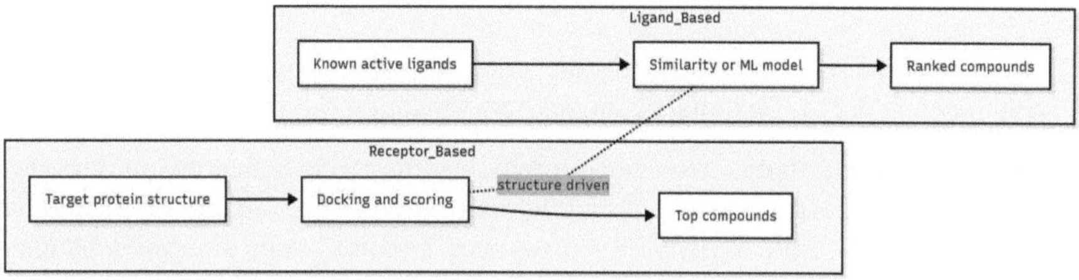


**Fig. 2.18** Receptor based vs Ligand based virtual screening

such as convolutional neural networks (CNNs) and recurrent neural networks (RNNs), can learn directly from raw molecular representations, including SMILES strings or molecular graphs. Variational autoencoders (VAEs) and generative adversarial networks (GANs) have been used to generate novel molecules with desirable properties, optimizing factors such as binding affinity, solubility, or toxicity.

In SBVS, the three-dimensional structure of a target protein is utilized to predict the binding affinity of small molecules. Traditional molecular docking methods can be slow and inaccurate. However, AI models, particularly graph-based neural networks, have demonstrated superior performance in approximating the complex relationships between a compound's structure and its binding behavior [33]. AI-driven protein structure prediction tools like AlphaFold have revolutionized SBVS by providing high-quality models of previously unresolved protein structures, thereby opening up new targets for drug discovery (Fig. 2.18).

One of the most promising applications of AI in personalized medicine is **predictive modeling for treatment outcomes**. This approach utilizes data-driven methods to predict a patient's response to a specific therapy, aiming to enhance clinical decision-making and tailor treatment plans to individual patients. Traditional treatment selection methods largely rely on population averages derived from clinical trials, which often overlook individual variability. AI enables a transition from a one-size-fits-all model to one that integrates genetic, clinical, behavioral, and environmental data, resulting in more accurate predictions.

Traditional clinical tools, such as **risk calculators and scoring systems,** provide some level of personalization; however, they are typically limited in scope and depend on a narrow range of variables. Moreover, these tools tend to be static and do not adapt as new patient data becomes available. In contrast, AI-enhanced predictive modeling enables dynamic, multivariable, and nonlinear interpretations of patient data [33]. It can continuously learn from new data and refine its predictions over time, potentially resulting in more accurate and context-aware treatment decisions.

Predictive models typically require the integration of various types of patient data, including:

- Electronic health records (EHRs): These contain longitudinal patient information, such as diagnoses, medications, lab results, and clinical notes.
- Genomics and transcriptomics: These provide molecular insights into disease susceptibility, drug metabolism, and potential therapeutic targets.
- Medical imaging: Radiomics and pathology imaging contribute quantitative features that can be correlated with outcomes.
- Wearable and sensor data: Real-time data on vital signs, activity levels, and adherence patterns can enhance model performance.

One of the most prominent areas where predictive modeling is making an impact is oncology. In this field, treatment decisions must be very precise and often involve balancing efficacy with toxicity. By analyzing genomic data alongside patient-specific clinical variables, AI models have been trained to predict how tumors will respond to various chemotherapy or immunotherapy agents. For instance, models trained on publicly available datasets, such as The Cancer Genome Atlas (TCGA), have demonstrated the ability to identify which patients are likely to respond to drugs based on their molecular signatures (Fig. 2.19).

Beyond oncology, predictive modeling applications have also been seen in cardiovascular medicine, where it can predict complications such as heart failure readmissions and arrhythmia risk. In psychiatry, AI is used to identify which patients are most likely to benefit from specific antidepressants, thereby reducing the trial-and-error phase often associated with psychiatric treatment. Predictive models have even shown utility in managing infectious diseases.

The concept of the **digital twin** has emerged as one of the most promising tools for bridging the complex realities of biology with intelligent computational modeling. In healthcare, a digital twin refers to a dynamic, continuously updated, and individualized digital replica of a patient. Essentially, it is a model that captures the patient's unique physiological, molecular, behavioral, and environmental characteristics, allowing for real-time simulation of disease progression and therapeutic interventions. The digital twin collects data from multiple sources, including electronic health records, wearable sensors, imaging modalities, genomics, and environmental influences, and integrates this information into a unified and computationally manageable representation. The model is not static; it continuously adapts as new

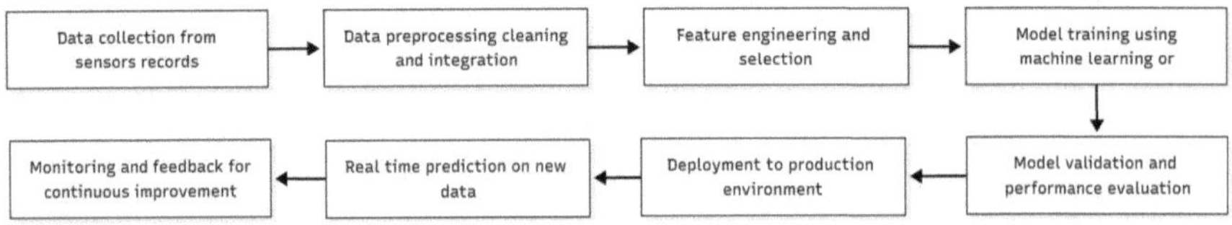


**Fig. 2.19** Predictive analytics

data becomes available, refining its predictions and improving its ability to forecast clinical outcomes. In this way, it evolves into a living, learning system that co-develops with the individual it represents [34].

Creating an effective digital twin requires the integration of extensive genomics and other multi-omics data across various spatial and temporal scales. In the field of cardiovascular medicine, for example, a digital twin could integrate hemodynamic simulations of blood flow with real-time ECG data and genetic risk assessments to predict the onset of arrhythmias or the response to treatments like beta-blockers. An innovative example of this is the **Digital Twin Heart,** developed by research groups in Europe and the U.S., which utilizes patient-specific imaging and electrophysiological data to create a real-time simulation of cardiac activity [35]. This model has been used to plan ablation procedures for arrhythmias and to virtually test responses to pacing strategies before they are implemented in patients (Table 2.3).

Artificial intelligence plays a crucial role in the operation and evolution of digital twins, particularly in predictive modeling. Key applications of AI include:

- Interpreting raw data inputs from sensors or clinical records
- Learning patterns of disease progression from historical data
- Predicting responses to interventions under various scenarios
- Updating the model as new data becomes available (continual learning)

**Table 2.3** Digital twins uses and applications

| Domain | Main Uses | Typical Applications | Key Benefits |
|---|---|---|---|
| Manufacturing and industry 4.0 | Real-time monitoring of machines and production lines | Predictive maintenance, process optimization, quality control | Reduced downtime, improved efficiency, cost savings |
| Healthcare and medicine | Personalized patient models for simulation and prediction | Virtual replicas of organs for surgery planning, disease progression modeling, drug testing | Better clinical decision making, individualized treatment, risk reduction |
| Smart cities and infrastructure | Representation of buildings, roads, and utilities | Energy management, traffic flow optimization, disaster preparedness | Improved urban planning, reduced energy consumption, faster emergency response |
| Energy and utilities | Simulation of power grids and renewable energy systems | Load balancing, predictive fault detection, grid optimization | Higher reliability, optimized resource use |
| Automotive and transportation | Vehicle performance modeling and fleet management | Autonomous vehicle testing, remote diagnostics, route optimization | Safer operations, faster design cycles |
| Aerospace and defense | Digital replicas of aircraft, satellites, and defense equipment | Structural health monitoring, mission simulation, predictive maintenance | Enhanced safety, lower operational costs |
| Retail and consumer services | Virtual models of stores and supply chains | Inventory optimization, customer experience simulation | Better demand forecasting, improved customer satisfaction |

In addition to predictive modeling, AI facilitates model calibration, ensuring that the digital twin remains aligned with real-world patient data. This involves solving optimization problems in real time and dynamically adjusting parameters to reflect changes in health status. Furthermore, natural language processing (NLP) techniques are increasingly employed to extract valuable insights from unstructured clinical notes and imaging reports. However, a significant concern is the computational complexity and scalability of these models. The processing power required to simulate high-fidelity models for millions of patients is substantial, raising important questions about cost, infrastructure, and feasibility in various healthcare settings.

## 2.5 Wearable Biosensors and Remote Health Monitoring

**Biosensors** are a newly described category of sensors that integrate physical and chemical sensing techniques. The International Union of Pure and Applied Chemistry (IUPAC) defines biosensors as chemical sensing analytical devices that convert chemical responses into electrical signals, offering high specificity and functioning independently of physical parameters such as temperature and pH. These biosensors utilize receptor-transducer tools to evaluate the biochemical and biophysical properties of a medium. A key feature of biosensors is their biological or organic recognition element, which enables the detection of specific biological substances. Common sensing elements include cells, antibodies, enzymes, and nucleic acids [36]. These components collaborate to create a recognition layer that is connected to the transducer through mechanisms such as cross-linking, covalent bonding, or adsorption.

**Wearable devices** can monitor and record data about a person's movements and physiological state in real time. Sensor-based health monitoring systems can incorporate various flexible sensors that can be directly attached to the human body or integrated into clothing, textile fibers, and elastic bands. These sensors measure physiological indicators such as blood pressure (BP), respiration rate (RR), body temperature, electrodermal activity (EDA), arterial oxygen saturation ($SpO_2$), heart rate (HR), electrocardiogram (ECG), and electromyogram (EMG). Additionally, they frequently employ gyroscopes, magnetic field sensors, accelerometers, and other miniature motion sensors based on micro-electro-mechanical systems (MEMS) to monitor activity-related signals [37]. Continuous monitoring of physiological signals can aid in the early detection and diagnosis of various neurological, pulmonary, and cardiovascular disorders. Moreover, tracking a person's movements in real time can assist in fall detection, posture and gait analysis, and sleep evaluation. Typically, wearable health monitoring systems include a combination of electrical and MEMS sensors, actuators, wireless communication modules, and signal processing units (Fig. 2.20).

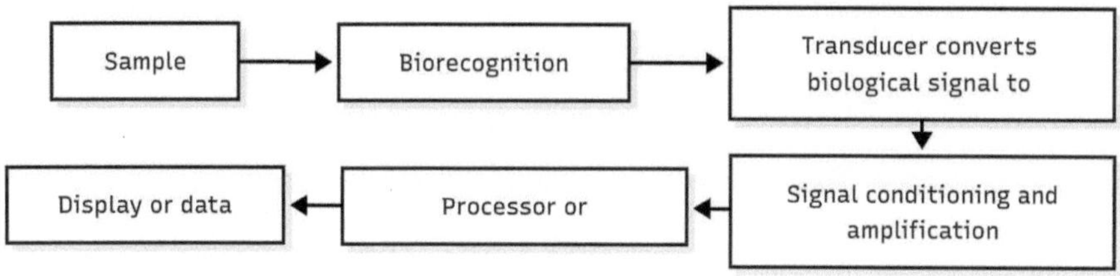


**Fig. 2.20** Block diagram of the structure of biosensors

### 2.5.1 *Types of Biosensors and Physiological Signals*

Biosensors are analytical tools designed to detect and measure specific physiological or biochemical signals. They achieve this by integrating a physical transducer with a biological recognition factor. In wearable and remote health monitoring systems, biosensors serve as the primary interface between the human body and digital health technology, enabling continuous assessment of biochemical markers and vital signs. Biosensors can be classified based on their form factor or platform, biological recognition element, and signal transduction process. Understanding these classifications is essential for designing systems tailored to specific healthcare applications.

**Electrochemical biosensors** are commonly based on enzymatic catalysis reactions that either produce or consume electrons. These enzymes, known as redox enzymes, are pivotal to their operation. Typically, these biosensors consist of three electrodes: a counter electrode, a reference electrode, and a working electrode. When the analyte interacts with the active electrode surface, it may facilitate electron transfer across a dual-layer potential, allowing the computation of current at a fixed potential. Electrochemical biosensors can be categorized into four types [38]:

1. **Amperometric Biosensors**: These measure the electric current generated during a biological reaction when an electroactive species is reduced or oxidized.
2. **Potentiometric Biosensors**: These do not require a current. Instead, they measure the change in electrical potential (voltage) between a reference electrode and a sensing electrode.
3. **Impedimetric Biosensors**: These monitor changes in electrical impedance (resistance and reactance) at the sensor interface, responding to biological interactions, such as when an antibody binds to an antigen. They utilize an alternating current (AC) signal to assess the analyte's impact on current flow.
4. **Voltammetric Biosensors**: These track the current response of an electrochemical system as the applied voltage changes over time. The target analyte undergoes redox (oxidation-reduction) processes at the electrode surface, producing a current that correlates with the analyte's concentration.
5. **Optical biosensors** use light-based methods—such as reflectance, fluorescence, or absorption—to track physiological changes. Photoplethysmography (PPG) is a key optical method employed in wearable devices to measure heart rate and

oxygen saturation ($SpO_2$) by assessing changes in blood volume within tissues. These sensors are commonly embedded in smartwatches and fitness bands.

Classification by biological recognition element:

1. **Enzyme-Based Biosensors:** These produce visible signals by catalyzing biological reactions using specific enzymes. They are often used to monitor metabolic indicators such as glucose, lactate, and urea due to their high specificity.
2. **Immuno Biosensors:** These use antibodies to detect specific antigens, such as viruses or biomarkers associated with diseases like cancer. Though not yet mainstream in wearables, this technology is rapidly advancing for applications such as COVID-19 detection.
3. **DNA and Nucleic Acid Biosensors:** These identify specific DNA or RNA sequences using complementary nucleic acid strands. Their primary applications lie in genetic research and diagnostics, with recent advancements pushing their use into wearable technology. They aim to analyze immune response variables through the quantification of DNA or antigen particles.
4. **Cell-Based Biosensors:** This type employs living cells as the detecting component. They leverage the capability of living cells to recognize physiological parameters and environmental conditions, producing reactions based on cellular interactions.

With these classifications, biosensors can be effectively utilized for various healthcare applications, paving the way for advancements in digital health technology. Sensors worn on the human body often detect weak signals that are susceptible to artifacts caused by human motion and external noise. The process of enhancing the target signal while eliminating noise and artifacts, followed by classifying and extracting distinctive signals for an analog-to-digital converter (ADC), is known as signal conditioning.

The signals collected by wearable sensors from the human body consist of analog impulses that are typically low in intensity, low in frequency, and narrow in bandwidth. Subsequent processing modules often struggle to handle these noisy, weak, and irregular inputs. To boost the amplitude of the raw signal, an amplification circuit is required. Amplification is a crucial component of wearable sensors and has been extensively researched. Transistors are commonly used to amplify the signal in wearable sensing circuits. One effective type of classical transistor used in biosensors is the **field effect transistor** (FET). A basic insulated gate FET consists of three terminals: source, gate, and drain, with an insulating $SiO_2$ layer separating the gate from the body. The flow of electrons and holes creates a depletion layer between the source and drain, and applying a voltage to the gate adjusts the amplitude of the drain current. The space between the source and drain in a wearable sensor can also function as a biometric component. Changes in analyte concentration lead to a shift in charge, which alters the gate voltage and subsequently changes the drain current, effectively converting and amplifying chemical signals into electrical signals.

To ensure that wearable technology conforms comfortably to the body's surface, components must often be lightweight, thin, and flexible. To meet these specifications, thinner components of the FET are used to create thin-film transistors (TFTs). Research has increasingly explored the potential of flexible sensors, including organic thin-film transistors (OTFTs) and organic electrochemical transistors (OECTs). Transistor amplifiers typically offer benefits such as low power consumption, compact size, and minimal power supply voltage requirements. They are also highly sensitive to voltage changes, resilient to physical shocks, and do not require preheating. TFTs, in particular, offer advantages like high processability and flexibility; OTFTs are flexible and printable, making them suitable for skin-surface sensors; and OECTs, which utilize liquid electrolytes, provide ultra-high flexibility and high transconductance at the expense of slower response times, making them ideal for amplifying low-frequency signals [39]. These technologies are widely used in wearable applications, including flexible electrocardiogram (ECG) patches, portable electroencephalogram (EEG) monitoring devices, and ultra-conformal drawn-on-skin electronics.

**Wearable sensors** are typically powered by small, low-voltage batteries. However, practical applications such as location tracking, healthcare, and environmental monitoring often require specific battery life, which raises questions about the ability of amplification circuits to operate efficiently at low voltages. Although OTFTs are quite flexible, they require higher driving voltages due to the low mobility and poor switching characteristics of organic semiconductors. Unfortunately, increasing the drive voltage also leads to higher losses in the amplifier circuit. Therefore, minimizing these losses involves reducing the drive voltage while maintaining a constant gain. Research into transistors with higher transconductance can effectively lower the driving voltage and, consequently, the power consumption. Various **noise sources** can negatively impact wearable sensors. For example, power line interference (such as 50 Hz or 60 Hz noise) can disrupt ECG signal acquisition, along with interference from motion artifacts and myoelectric clutter. To eliminate noise while preserving the signal of interest, the choice of filtering method must be tailored to the noise characteristics and filtering requirements [40]. Filters are classified into analog and digital categories based on the type of signal they are designed to process. A simple analog filter can be constructed by utilizing the properties of capacitors to block low-frequency signals and inductors to allow high-frequency signals to pass through. By adjusting the size and arrangement of the inductor and capacitor, the analog filter can be designed to let only signals within a specific frequency band pass. On the other hand, digital filters comprise an adder, a digital multiplier, and a delay unit. Unlike analog filters, digital filters alter the digital code of the input discrete signal to modify its frequency, rather than relying on the electrical properties of components for filtering [41]. Digital filters tend to be more intricately designed and smaller than their analog counterparts. Common filters used in wearable sensors can be categorized into three types based on their frequency response functions.

### 2.5.2 AI for Anomaly Detection and Risk Prediction

Since the 1960s, the detection of anomalies, also known as outlier or novelty detection, has been a continuous area of study across various application sectors. The advancement of computer processes has facilitated the gathering of big data and its application in artificial intelligence (AI), particularly in time series analysis and anomaly identification. With the increased availability of data and enhancements in algorithmic efficiency and computational capacity, time series analysis is increasingly being employed to tackle business applications through forecasting, classification, and anomaly detection. This growing significance of time series anomaly detection (TSAD) spans a wide range of fields, including urban management, intrusion detection, medical risk assessment, and natural disaster response.

Anomalies can be categorized into three types: temporal, intermetric, and temporal-intermetric. **Temporal anomalies** in a time series can be further classified as local, comparing them to their neighboring values, or global, comparing them to the entire time series. These anomalies are often defined by the difference between the observed and expected subsequences. A useful technique for quantifying this dissimilarity is dynamic time warping (DTW), which optimally aligns two time series [42]. Therefore, DTW is an effective tool for anomaly detection applications due to its precision in calculating temporal alignments.

In practice, training data typically contains very few labeled anomalies. As a result, most models focus on learning the properties or representations of normal data. The next step in detecting anomalies involves identifying deviations from this typical data. Recent deep learning models for anomaly detection employ four different learning schemes: self-supervised, semi-supervised, supervised, and unsupervised, depending on the availability of labeled data points. The supervised approach utilizes all the labels in the training set to differentiate between anomalous and normal data. If the anomaly score assigned to certain timestamps exceeds a threshold, it identifies an appropriate threshold value to categorize all timestamps as anomalous. However, this method has the downside of being unsuitable for many real-world applications due to the frequent lack of labeled anomalies.

In contrast, the unsupervised strategy does not differentiate between training and testing datasets and does not rely on labels. These methods are more adaptable because they utilize the inherent characteristics of the data without needing labels for testing or training, making them particularly useful for streaming applications. Nonetheless, researchers may face challenges when evaluating anomaly detection models using unsupervised techniques, as historical data is typically unlabeled, and anomalies can be unpredictable [43]. Semi-supervised anomaly detection in time series data can be applied when the dataset contains labeled normal data. This method leverages labeled normal data to describe normal patterns and identify deviations as anomalies, differing from unsupervised approaches, which detect anomalies without any labeled data. Furthermore, self-supervised learning enables models to generate their own supervisory signal from the input data, eliminating the need for explicit labels.

### 2.5.3 Edge Computing and Real-Time Analytics

Wearable technology can monitor various bodily symptoms, such as heart rate, breathing, movement, and brain activity. These small devices utilize a range of sensors to detect, predict, and analyze the physical performance, physiological state, biochemical makeup, and mental awareness of the human body. Despite advancements in new materials that could enhance sensor sensitivity and resolution, modern wearable technology faces several challenges, including low processing power, high energy consumption, large data transmission volumes, and slow data transfer speeds. Typically, the data collected by traditional wearable sensing solutions is sent to external servers for processing and computation [44]. This approach often leads to information bottlenecks, which hinder efforts to reduce power consumption and improve the speed of sensing systems. Additionally, processing real-time sensing data typically relies on conventional remote servers using traditional signal processing methods, resulting in increased computational load and significant hardware requirements.

In this context, the **edge computing paradigm** has gained significant attention. This approach encompasses any network in which the processing node is not located in the cloud. Generally, power efficiency improves the closer the sensing and computation units are to each other. Specifically, when data processing occurs on the same device as the sensor, it is referred to as "extreme edge computing." This paradigm shift necessitates a new perspective. While general-purpose systems are powerful and flexible, they do not account for the diverse volume and quality of data generated by various devices. Understanding this, a tailored solution that leverages available resources for specific tasks may be more advantageous than a general-purpose system in terms of power efficiency, area efficiency, and latency. Moreover, typical processing units might not effectively tackle the aforementioned challenges, even when computing is pushed to its limits. In standard von-Neumann systems, the memory and processing units are physically distinct, which increases the power needed to transfer data between them. Consequently, there is usually a trade-off between computational capacity and power lifespan in such systems. By performing computation at the edge, we can achieve faster response times and enable customized wearable technology that continuously interacts with and learns from its environment. To optimize system performance while minimizing power consumption and memory requirements, a significant paradigm shift that incorporates cutting-edge algorithms, circuits, and memory devices is essential (Table 2.4).

Fuzzy inputs and complex cognitive tasks, such as pattern recognition, time series prediction, and decision-making, are not well-suited for conventional computers, which rely on Boolean and bit-precise digital representations and operate through time-multiplexed and clocked signals. In contrast, deep artificial neural networks (ANNs) have demonstrated impressive performance in various pattern recognition tasks, including speech recognition, machine vision, and natural language processing (NLP). To execute these algorithms and produce highly accurate inference results, accelerators for Artificial Neural Networks (ANNs), including GPUs,

**Table 2.4** Wearable biomedical signals and sensors

| Biomedical Signal | Typical Sensor Type | Measurement Site | Common Applications | Key Considerations |
|---|---|---|---|---|
| Electrocardiogram (ECG) | Dry or wet electrodes, textile electrodes | Chest, wrist, ankle | Cardiac rhythm monitoring, arrhythmia detection, heart rate variability | Motion artifacts, electrode-skin contact quality |
| Photoplethysmogram (PPG) | Optical sensor with LED and photodiode | Finger, wrist, ear lobe | Heart rate, blood oxygen saturation ($SpO_2$), blood pressure estimation | Sensitive to ambient light and movement |
| Electroencephalogram (EEG) | Dry or gel-based electrodes | Scalp | Brain-computer interfaces, seizure detection, sleep studies | Requires good electrode placement, prone to noise |
| Electromyogram (EMG) | Surface electrodes | Forearm, leg, facial muscles | Prosthetic control, muscle fatigue assessment | Cross-talk from adjacent muscles, motion artifacts |
| Electrodermal activity (EDA/GSR) | Skin conductance electrodes | Fingers, palm, wrist | Stress monitoring, emotion recognition | Affected by temperature and sweat |
| Body temperature | Thermistor, infrared sensor | Wrist, chest, ear | Fever detection, circadian rhythm monitoring | Sensor placement and ambient temperature influence accuracy |
| Respiration rate | Stretch sensor, piezoelectric belt, accelerometer | Chest, abdomen | Sleep monitoring, respiratory disease tracking | Motion interference, calibration required |
| Blood pressure (cuffless) | PPG plus ECG or tonometry | Wrist, finger, upper arm | Continuous hypertension monitoring | Requires calibration against cuff-based measurements |
| Blood glucose (non-invasive prototypes) | Optical spectroscopy, electrochemical patch | Upper arm, wrist | Diabetes management | Accuracy still limited compared to invasive methods |
| Motion/activity | Accelerometer, gyroscope, magnetometer | Wrist, ankle, chest | Step counting, fall detection, energy expenditure | Drift and sensor fusion errors |

TPUs, and custom ASICs with parallel architectures, are under development. Notably, the extended memory bus of these accelerators is particularly effective for running vector-matrix multiplications (VMMs), which are central to the processing in deep neural networks [45]. While GPUs provide a parallel processing foundation, offering some parallelism, it is on a much smaller scale than the brain. However, due to their dependence on timing and the non-localized nature of memory access, they consume significantly more power than the brain.

To address this issue, **Application-Specific Integrated Circuit (ASIC)** accelerators aim to simplify system architecture by increasing application specificity. They employ techniques such as clock gating and specialized hardware that aligns with the structure of the mapped neural network, ultimately reducing power consumption through fewer memory reads and data accesses. Two critical issues must be resolved to further decrease power consumption: (i) eliminating the clock, and (ii) executing computations with co-located memory and processing units. To overcome the first issue, developers must create event-based systems that process information "asynchronously," or only when input "events" occur. The Spiking Neural Network (SNN), in which neurons communicate asynchronously through spikes, serves as the algorithmic foundation for this "asynchronous" processing.

Additionally, the memory element should be utilized for both computation and data storage within the processor to minimize data movement between memory and processing units. This approach is known as "in-memory computing."

The **advantages of wearable biosensors have made real-time monitoring** of vital signs easier, which is essential for fitness and health management. These devices provide non-invasive, immediate results, allowing users to assess their physiological conditions and make informed healthcare decisions. Applications of such monitoring include tracking vital signs such as blood pressure, body temperature, oxygen saturation, heart rate, glucose levels, physical activity, and sleep patterns.

**Continuous glucose monitors (CGMs)**, which are worn on the body, offer real-time blood sugar measurements and are particularly beneficial for individuals with diabetes. These devices enable users to better manage their insulin doses and dietary choices by measuring glucose levels in intestinal fluid, thereby reducing the risk of hypoglycemia. Smartwatches and fitness trackers assess a user's heart rate using electrocardiography (ECG) or photoplethysmography (PPG). They provide real-time summaries of heart rate variability, helping users monitor changes in their cardiovascular health, maintain physical activity, and identify potential arrhythmias and other heart-related issues. Additionally, pulse oximeters, which monitor oxygen saturation using light absorption techniques, are crucial for individuals with respiratory conditions such as asthma or chronic obstructive pulmonary disease (COPD), allowing them to continuously check their oxygen levels (Fig. 2.21).

Recent developments in wearable technology for blood pressure monitoring eliminate the need for traditional cuff-style equipment, enabling users to easily assess their heart health. Continuous blood pressure monitoring can detect hypertension and encourage individuals to seek medical advice sooner for targeted treatment. Wearable biosensors can also track body temperature in real-time, aiding in

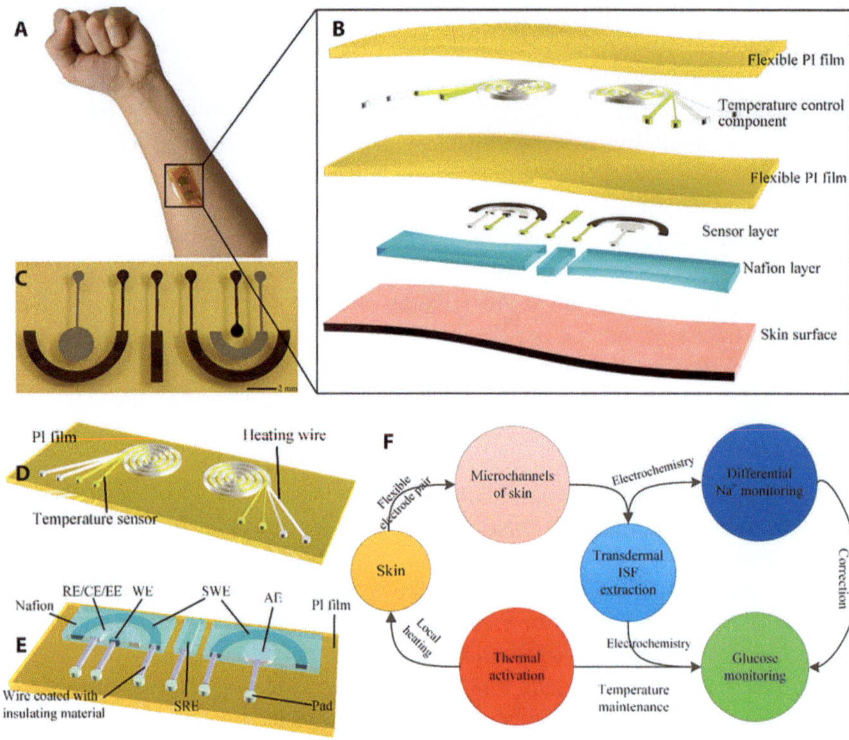


**Fig. 2.21** Continuous glucose monitoring (https://commons.wikimedia.org/wiki/File:Design_of_the_flexible_epidermal_biomicrofluidic_device_for_continuous_blood_glucose_monitoring.jpg)

the detection of fevers and other medical conditions [46]. Continuous temperature data can reveal health trends and assist in managing common illnesses. Furthermore, some wearable sensors can capture movement and sleep patterns, providing valuable insights into overall health beyond vital signs.

By utilizing these devices, individuals can modify their lifestyles to enhance bodily performance and monitor exercise levels and sleep quality. Overall, wearable biosensors simplify health management by continuously displaying key health indicators such as blood oxygen levels, heart rate, and blood sugar levels, empowering individuals to take control of their health.

## 2.6 Robotics and AI in Surgery and Rehabilitation

Although the roots of robotic surgery trace back to the 1980s, the twenty-first century marks a significant era for robotics and artificial intelligence (AI) as they dramatically reshape healthcare alongside many other fields. Their impact is particularly evident in surgery and rehabilitation, where they enhance precision during operations and improve recovery efficiency like never before.

Surgical procedures have traditionally relied on human knowledge, extensive experience, and specialized skills. Rehabilitation often involved repetitive exercises. However, with the advent of advanced robots and intelligent software, the medical landscape is undergoing a profound transformation, becoming more precise, efficient, and personalized. Robotic systems provide surgeons with extra capabilities, enabling them to perform complex procedures that require fine motor skills and exceptional precision [47]. Nevertheless, the role of robots extends beyond simple task execution. Planning, predicting potential complications, and guiding both robotic and human efforts fall within the realm of artificial intelligence.

**AI-powered robots** are being utilized in physical therapy to assist patients with exercises, monitor their movements, and develop customized recovery plans. The advancement of artificial limbs (prosthetics) and wearable robotic suits (exoskeletons) allows individuals with disabilities to regain independence and control, thereby enhancing the likelihood of successful rehabilitation. AI plays a crucial role by analyzing vast amounts of data and functioning as an efficient assistant that never overlooks a detail. Understanding the influences of robots and AI on surgery and rehabilitation emphasizes the importance of developing and continually improving the human-robot relationship. At the same time, it is critical to ensure that these technologies are utilized ethically and wisely.

The origins of robotic surgery can be traced to the 1980s, when military interest in developing remote surgical options for battlefield injuries prompted early advancements. A significant milestone was reached in 1985 with the introduction of the PUMA 200, a robotic arm used for neurosurgical biopsies [48]. The following decade saw the emergence of more specialized systems, such as the PROBOT in 1992 for prostate surgery and the ROBODOC in the same year for hip replacement procedures. In 1997, the AESOP (Automated Endoscopic System for Optimal Positioning) system was introduced, allowing surgeons to control the camera using voice commands, thereby enhancing control during laparoscopic operations (Fig. 2.22).

The most substantial advancement occurred in the 2000s with the FDA approval of the da Vinci Surgical System, which quickly established itself as the leading platform for a variety of minimally invasive surgeries, offering high precision and enhanced 3D visualization. Since then, the field has continued to grow rapidly, with the introduction of new robotic platforms and AI systems for planning and navigation, along with an increasingly diverse range of surgical applications.

**Robot-assisted surgery,** also known as robotic surgery, marks a significant advancement in modern medicine. This innovative approach provides a less invasive alternative to traditional open surgery and conventional laparoscopic techniques. It allows surgeons to perform complex operations with enhanced precision, flexibility, and control, often through very small incisions. In a typical robot-assisted surgical setup, the surgeon does not directly manipulate the surgical instruments. Instead, they sit at a console located in the same operating room, controlling robotic arms that are equipped with specialized surgical tools and a high-definition 3D

**Fig. 2.22** Puma 200 (https://commons.wikimedia.org/wiki/File:UNIMATE_PUMA_200_Robot_Arm_(6202074072).jpg)

camera [49]. The robotic system translates the surgeon's hand movements into precise, scaled movements of the instruments inside the patient's body. The 3D camera offers a magnified, high-resolution view of the surgical site, enabling the surgeon to see intricate details that may be difficult to recognize with the naked eye. Popular systems like the da Vinci Surgical System allow surgeons to perform procedures with remarkable dexterity and control. Other systems, such as those developed by Medtronic, also contribute to the growing field of surgical robotics (Fig. 2.23).

Robot-assisted surgery offers numerous benefits for both patients and surgeons. For patients, these procedures typically involve smaller incisions, resulting in less scarring and reduced postoperative pain. Compared to traditional open surgery, there is significantly less blood loss, shorter recovery times, and shorter hospital stays, all of which contribute to a lower risk of infection. Consequently, patients can return to their normal activities more quickly.

From the surgeon's perspective, robot-assisted surgical systems greatly enhance the ability to operate with extreme accuracy and skill. These robots enable a wider range of movement and provide more delicate control. The magnified 3D

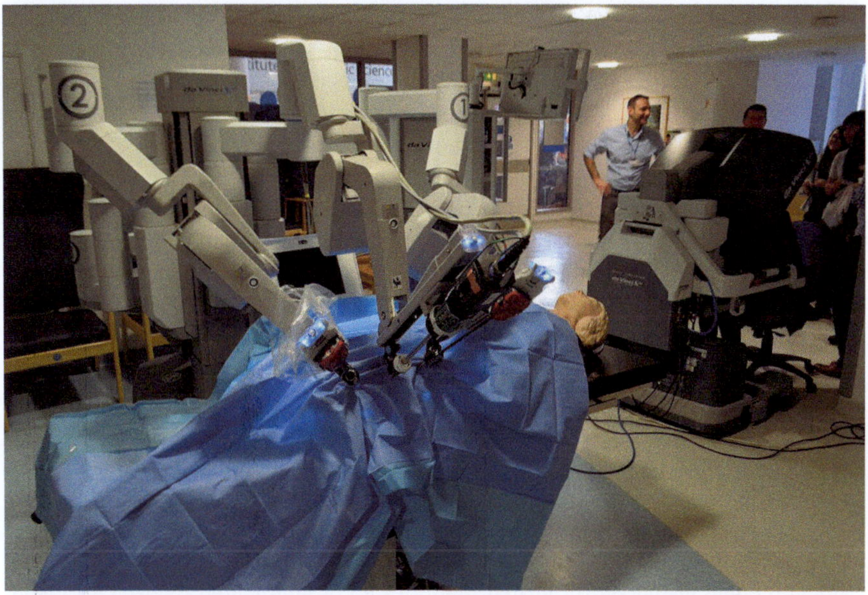

**Fig. 2.23** Robot assisted surgery (https://commons.wikimedia.org/wiki/File:Cmglee_Cambridge_Science_Festival_2015_da_Vinci.jpg)

high-definition vision included in the system improves depth perception and clarity of the surgical field, while tremor filtration ensures steady movements [50]. This is crucial for precise cutting and stitching, especially in hard-to-reach areas of the body. Moreover, the ergonomic design of these machines enables surgeons to operate from a comfortable seated position at the console, thereby reducing the physical strain that often accompanies prolonged traditional procedures.

Robot-assisted surgery is utilized across various medical specialties, including urology, gynecology, colorectal, cardiac, general, spine surgery, and neurosurgery. In urology, it is used for procedures such as prostatectomy (removal of the prostate gland) and kidney surgery, where precision is essential for preserving nerves and minimizing complications. In gynecology, it can be employed for hysterectomy (removal of the uterus), myomectomy (removal of uterine fibroids), and other pelvic surgeries. This technique is also applicable in colorectal surgeries for removing colon and rectal cancers, in cardiac surgery for mitral valve repair and coronary artery bypass, as well as in general surgery for hernia repair, gallbladder removal, and numerous other procedures.

This modern technology continues to evolve. Ongoing research holds promise for its application in more complex and delicate procedures, such as transplant surgeries and oncological treatments. While it currently offers significant advantages, it is essential to note that it may not be suitable for all patients or types of procedures. The decision to use this technique is made by the surgical team based on the specific needs of each patient [51]. Despite its many significant advantages, it is important to recognize that, like any surgical procedure, robot-assisted surgery carries potential risks. These include adverse reactions to anesthesia, bleeding, infection at the surgical site, and the possibility of damaging surrounding nerves, organs, or tissues. In rare cases, technical malfunctions may occur; however, these systems are designed with extensive safety protocols to minimize such risks.

AI is increasingly playing a crucial role in optimizing surgical strategies by significantly enhancing both **preoperative planning and intraoperative navigation.** Before any incision is made, AI systems can process vast amounts of medical data, enabling surgeons to perform procedures that are more precise, efficient, and safe. One of the primary applications of AI in this context is advanced image analysis. AI algorithms, particularly those based on deep learning, can analyze medical images, such as CT scans, MRIs, and X-rays, with remarkable speed and accuracy, often surpassing the capabilities of some medical specialists in the field. These algorithms can detect and identify abnormalities that might be missed by the human eye by accurately segmenting organs and tumors, as well as mapping complex anatomical structures in high-resolution 3D imaging. For instance, in oncology, AI can accurately delineate tumor boundaries, assisting in the complete resection of tumors while preserving healthy tissue. Additionally, AI can create patient-specific anatomical models visualized in virtual reality (VR) or augmented reality (AR) environments [52]. This technology can be utilized by surgeons for training, enabling them to visualize the surgical field in advance, identify potential challenges, simulate various approaches, and practice complex movements. By analyzing extensive patient data, including medical history, genetic information, laboratory results, and imaging findings, AI assists in risk assessment and predicting outcomes. It identifies patterns, predicts complications, estimates recovery times, and suggests the most suitable surgical approaches tailored to each patient. This data-driven decision-making process leads to personalized care plans, helping to address potential issues and ultimately improving overall surgical success rates while enhancing patient safety.

AI-powered navigation systems provide real-time guidance, serving as an intelligent co-pilot for surgeons. These systems integrate preoperative plans with **live intraoperative data** from endoscopic cameras, ultrasound, and specialized electromagnetic tracking sensors. The goal is to accurately track the positions of surgical instruments in relation to the patient's anatomy. AI navigation systems can identify nerves, major blood vessels, and delicate organs that may not be visible directly. If the surgeon approaches a high-risk area or deviates from the planned route, the AI system provides immediate visual or auditory alerts. This real-time feedback significantly enhances the surgeon's situational awareness and reduces the risk of accidental injuries. The integration of augmented reality (AR) powered by AI represents another significant advancement in surgical navigation. AR systems overlay digital information—such as 3D anatomical models, tumor boundaries, or surgical pathways—directly onto the patient's body or the surgeon's field of view using specialized headsets or monitors [53]. This allows surgeons to "see through" tissues and precisely target areas of interest, particularly in delicate or minimally invasive procedures where direct visualization is limited. By analyzing live physiological data and suggesting optimal next steps, augmented reality also contributes to intraoperative decision support (Fig. 2.24).

**Exoskeletons** are wearable robotic devices designed to assist individuals in improving their joint movement. Unlike prosthetics, which replace lost body parts, exoskeletons are worn externally by individuals who still have their limbs but require assistance with movement. This includes people recovering from strokes, spinal cord injuries, or those facing age-related muscle and bone issues. These systems can range from simple designs that provide passive support to more advanced versions equipped with motors, sensors, and small computers that actively assist in walking and other movements.

Several lower-limb exoskeletons have received FDA approval for use in rehabilitation, including EksoNR (Ekso Bionics), ReWalk, and HAL (Hybrid Assistive Limb by Cyberdyne). These devices help patients regain their ability to walk by supporting hip and knee movements. They utilize motion sensors, force sensors, and muscle signals (EMG) to interpret a user's intentions and provide synchronized motor assistance. Regular use of exoskeletons in gait training can strengthen nerve connections, improve cardiovascular health, and reduce muscle stiffness in patients with spinal cord injuries.

Exoskeleton technology also has important **applications in industrial settings**, where it protects workers and reduces physical strain. Devices like the Guardian XO by Sarcos Robotics and the HeroWear Apex are back-assist exosuits designed to alleviate stress on the lower back during lifting. These tools are particularly valuable in logistics, construction, and manufacturing, as they help prevent common musculoskeletal injuries resulting from repetitive lifting. The U.S. military is also invested in exoskeleton research to enhance soldier endurance [54]. The Soldier Assistive Bionic Exosuit for Resupply (SABER), developed in collaboration with Vanderbilt University and the U.S. Army, helps soldiers carry loads during long patrols. This design has evolved from early combat-focused models to prioritize logistics and injury prevention (Fig. 2.25 and 2.26).

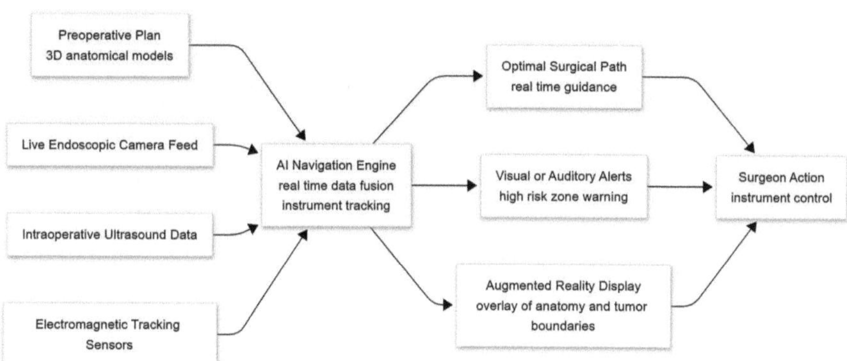


**Fig. 2.24** AI navigation engine

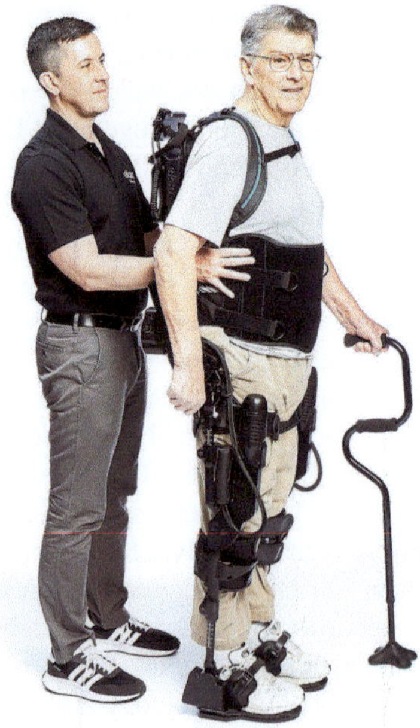

**Fig. 2.25** EksoNR (https://commons.wikimedia.org/wiki/File:EksoNR.jpg)

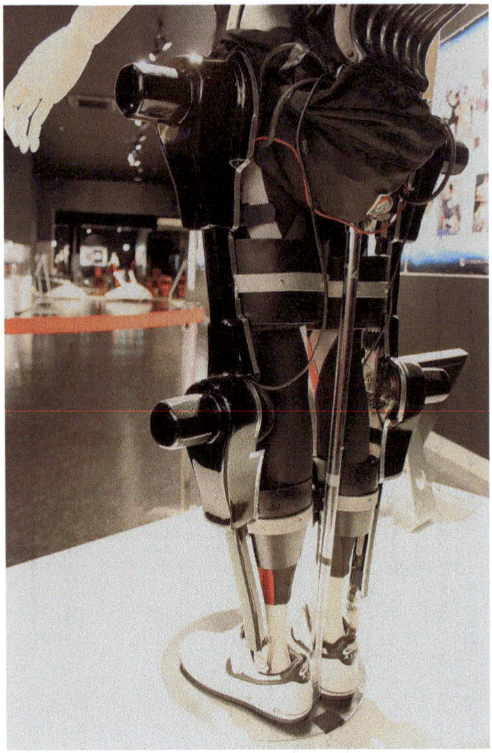

**Fig. 2.26** Exoskeleton (https://commons.wikimedia.org/wiki/File:Hybrid_Assistive_Limb,_CYBERDYNE.jpg)

Despite their potential, smart prosthetics and exoskeletons face several challenges. Factors such as device weight, comfort, and ease of wearing significantly influence user adoption. Adapting to varying walking surfaces, like stairs or uneven ground, requires sophisticated real-time control. Researchers are focusing on developing lighter materials, modular designs, and intelligent control systems to address these challenges. The ultimate goal is to create adaptive systems that can accurately respond to user intentions and changing environments, moving beyond simple muscle signal detection or pre-programmed movements.

### *2.6.1 Brain-Computer Interfaces and Neural Integration*

**Assistive devices** are increasingly utilizing direct brain control through **Brain-Computer Interfaces (BCIs).** Non-invasive BCI systems, such as those based on electroencephalography (EEG), can detect user intent but often struggle with signal noise. In contrast, invasive neural interfaces, like those used in the BrainGate project, provide higher-resolution decoding of brain activity. This enables patients with paralysis to control robotic arms or computer cursors using their thoughts. Prosthetic devices are also advancing, incorporating osseointegration and implanted electromyography (EMG) electrodes for stable and long-term neural signal capture. For instance, a neuromusculoskeletal arm prosthesis allowed a patient to experience sensory feedback and precise motor control for three consecutive years without any issues. Looking ahead, soft robotics and biohybrid systems are pushing the boundaries of what is possible. Harvard's Wyss Institute is developing soft exosuits made from textiles and cable-driven mechanisms. These suits are lightweight, flexible, and move more naturally with the body compared to rigid exoskeletons, providing discreet assistance for walking and running.

**Rehabilitation robotics** and motion analysis play a crucial role in modern therapy for conditions affecting movement. These systems assess, monitor, and enhance recovery through data-driven sessions that extend beyond traditional methods. Devices such as the Lokomat and InMotion ARM utilize robotics, biomechanics, and AI to deliver personalized therapy along with feedback in virtual environments, thereby enhancing patient engagement and outcomes [55]. Motion analysis tools, including optical sensors and wearable devices, track patient progress and refine treatment plans. AI assists in analyzing this data to identify abnormal movements or predict fall risks. Advanced systems, like the Motek CAREN, even simulate challenging environments in virtual reality to train balance and adaptive control. Robotic rehabilitation supports the brain's ability to rewire itself, a phenomenon known as neuroplasticity. Studies indicate that repeated robot-assisted therapy increases brain activity in damaged areas, aiding patients in relearning movement. For instance, research involving the InMotion ARM demonstrated significant improvements over traditional therapy within 12 weeks, primarily due to its capacity to adjust resistance and provide precise feedback. Other robotic platforms, such as ArmeoSpring and ReoGo, are also used in clinics to assist with arm and hand movements. These systems offer adjustable resistance and virtual environments to keep patients engaged and improve outcomes. AI is essential for personalizing rehabilitation robotics. AI models learn from patient performance over time, allowing them to dynamically adjust therapy difficulty, select optimal movement paths, and predict recovery outcomes. This adaptability enables robots to modify their assistance strategies in response to a user's real-time feedback and improvements in motor skills. A study involving stroke patients found that adaptive, AI-driven upper limb therapy programs resulted in a 25% greater improvement compared to fixed programs [56]. The AI could adjust assistance levels, the frequency of repetitions, and even the type of feedback in real-time based on each patient's fatigue, performance, and engagement. This underscores the importance of accurate motion analysis—now achievable with wearable AI platforms—in diagnosing movement issues, tailoring interventions, and tracking progress throughout rehabilitation.

**Closed-loop systems** continuously sense, interpret, actuate, and adapt. They utilize biosignals, such as EMG and EEG, along with movement data and patient feedback to adjust therapy in real-time. For example, detecting effort through muscle signals can adjust robot resistance, while brain signals can alter task difficulty. These systems enhance engagement and outcomes. The RecoveriX system utilizes brain signals (EEG) to control robotic arm movements for stroke patients, leading to increased brain activity and voluntary movement after 8 weeks of brain-computer interface (BCI) therapy.

Integrating various data types, such as movement patterns, heart rate, and fatigue levels, enables precise rehabilitation. AI-powered systems analyze this data to create personalized motor profiles and adapt therapy plans accordingly [57]. Algorithms like reinforcement learning and Bayesian optimization excel in this area, dynamically selecting the most suitable exercises and intensity. Trials demonstrate that these adaptive systems outperform traditional static therapy.

**Tele-rehabilitation platforms** are emerging that combine remote robotic devices, wearable sensors or video for motion capture, and cloud-based AI. This approach allows therapists to remotely monitor progress, update programs, and deliver interventions to stroke survivors at home. For example, the **HARMONY Tele-rehab System**, which includes an arm trainer and inertial measurement unit (IMU), has demonstrated results comparable to in-clinic therapy for post-acute stroke patients [58]. This accessibility is especially vital for those in rural areas, the elderly, or individuals with limited mobility, as it eliminates geographical and logistical barriers to consistent therapy.

## 2.7 AI in Biomedicine: Ethical, Regulatory, and Future Considerations

The use of AI in biomedicine has led to significant advancements, including the development of new drugs, earlier disease diagnoses, and personalized patient treatments. However, these benefits raise serious privacy concerns, as these developments heavily rely on sensitive patient data. Tools employing AI in biomedicine, such as those for analyzing medical images, studying genetic information, and predicting health risks, require access to detailed patient records. This often includes personal information, which creates substantial privacy and security risks [59]. These tools frequently collect patient information continuously and in real-time, making it essential to understand how data is collected and stored. Once information is provided to an AI system, it cannot be withdrawn; therefore, obtaining patient consent and agreement for the data to be included in the system is crucial.

Currently, there are risks that data could be:

- Leaked
- Accessed without permission
- That patients might be identified from information meant to be anonymous.

If a data breach occurs, it can lead to:

- Identity theft
- Discrimination
- A loss of trust in healthcare providers
- Serious financial or legal problems for the organizations responsible.

This is why data protection in AI is vital, as it directly affects patient safety and privacy.

**Protected Health Information (PHI)** contains highly sensitive personal details that go beyond typical identifiers like names and addresses. It includes information about a person's medical history, diagnosis, treatment plans, prescriptions, mental health records, genetics, biometrics (such as fingerprints or retinal scans), test results, and any other information linked to their medical care. PHI serves many important purposes. For doctors and patients, accurate diagnosis and proper treatment are essential. For medical researchers, anonymized PHI facilitates scientific progress. Due to its potential use in identity theft, insurance fraud, illegal prescription scams, bribery, or sales on the dark web, criminals often view PHI as more valuable than credit card information. Unlike financial data, healthcare records contain permanent personal information that is much harder to replace once stolen. Hospitals utilize PHI when applying AI to analyze medical images or predict health risks based on genetic data, which enables physicians to provide better, more personalized care [60]. Because PHI encompasses permanent information like genetic details or mental health records, it is significantly more challenging to modify or rectify if stolen compared to items like credit card numbers. When AI systems handle health information, they must comply with specific laws designed to protect patient privacy. The HIPAA is the primary federal law safeguarding health information in the United States. It mandates measures to secure all forms of PHI and establishes national standards for the management of electronic health records. For instance, if a hospital employs an AI tool to analyze X-rays for signs of cancer, that system must be secured by passwords, with access limited to authorized personnel. HIPAA also requires that data be properly concealed and backed up to prevent loss or unauthorized disclosure.

The **Health Information Technology for Economic and Clinical Health Act (HITECH)** expands on HIPAA by enhancing rules related to health information technology, data security, and breach notifications. For example, if a medical app controlled by AI mistakenly sends patient results to the wrong email, under HITECH, the app company or hospital is obligated to report the error and rectify it. The GDPR, the EU's data protection law, categorizes health information as sensitive and imposes strict regulations on its handling and processing [61]. If a company is training an AI system to predict diseases based on symptoms entered into a phone app, it must first obtain user consent. The company can only retain that data for as long as necessary and must clearly inform users how it will be used. Additionally, most U.S. states have implemented their own privacy laws, often providing extra protections specifically for health data beyond federal requirements. For example, in California, if an AI tool is used to track an individual's therapy sessions or mood patterns, the data must be stored with enhanced security measures, and the patient must have greater control over who can access it.

**Artificial intelligence bias** (AI bias) occurs when AI systems are constructed in an unfair manner that results in discrimination. This type of bias can reinforce existing stereotypes and cause harm by perpetuating or exacerbating the unfair treatment of certain groups.

Bias in AI generally arises from two primary sources:

1. The design of the models, and
2. The data used for training.

At times, the creators of AI models incorporate their own assumptions, which can lead the system to favor certain outcomes over others. AI bias can also stem from the training data itself. AI models learn by analyzing large datasets through a process known as machine learning. They identify patterns and connections within this data to aid in decision-making and predictions. If the data reflects historical biases or unfair disparities, the AI will likely adopt and reproduce these biases in its outputs. Even minor biases in the training data can lead to significant negative effects, as AI systems process vast amounts of information [62]. Biased training data can lead to discriminatory treatment for specific groups in fields such as biomedicine, impacting risk prediction, treatment recommendations, and disease diagnosis. AI systems may fail to deliver accurate results for individuals from diverse backgrounds if the majority of training data originates from a single ethnic group. This can lead to incorrect diagnoses or inappropriate treatments, particularly in areas where demographic differences are critical, such as genetic testing or dermatology. The consequences can be detrimental and may perpetuate existing issues. Additionally, bias reduces the accuracy of AI, limiting its potential to truly assist. Companies do not benefit from AI systems that yield biased results [63]. When the public hears about AI causing problems due to bias, trust in these systems may erode, particularly among marginalized groups such as people of color, women, individuals with disabilities, and LGBTQ+ individuals.

Ensuring **equitable access to care for everyone,** regardless of their background, is a cornerstone of healthcare fairness. This encompasses who receives particular treatments, how medical resources are allocated, and the outcomes that individuals experience. It is founded on fundamental ethical principles such as fairness, beneficence, and non-maleficence. A fair healthcare system should not differentiate treatment based on factors like income, gender, or race. Fairness in radiology AI refers to the technology's capacity to perform equally for all patients [64]. No group should receive better outcomes while another group receives worse ones. Achieving this requires a comprehensive understanding of the sources of bias in AI systems and the development of strategies to mitigate them.

In healthcare, **biases in AI systems** typically originate from two primary sources: the data used for training, which may already contain inherent biases, and the design of the algorithms or their adaptive learning processes. These are commonly referred to as data bias and algorithmic bias. However, healthcare differs from other fields; it involves complex human interactions, introducing additional layers of bias. Some extra biases arise from how healthcare professionals, such as doctors and nurses, engage with AI tools. Others occur when patients interact with these systems. For instance, if a doctor relies too heavily on AI recommendations or disregards them entirely, it can adversely affect patient care [65]. Additionally, if a patient distrusts AI or feels excluded from the decision-making process, it can undermine the effectiveness of treatment. These forms of interaction introduce further avenues for bias and should be considered just as significant (Fig. 2.27).

**Accountability and transparency** are crucial for the ethical use of AI in healthcare. They foster trust in the system by helping both healthcare professionals and patients comprehend how AI makes its predictions. To enhance accountability, it is crucial to develop algorithms that are transparent, clearly demonstrate the decision-making process, and provide clear justifications for AI's predictions. When AI is more transparent, it has the potential to improve decision-making for both patients and doctors. Additionally, transparency helps address moral and legal concerns related to AI applications in healthcare. However, it is crucial to acknowledge the limitations of clarity. Saliency maps can indicate which elements of an image influenced the AI's decisions, but it is still up to humans to interpret the significance of

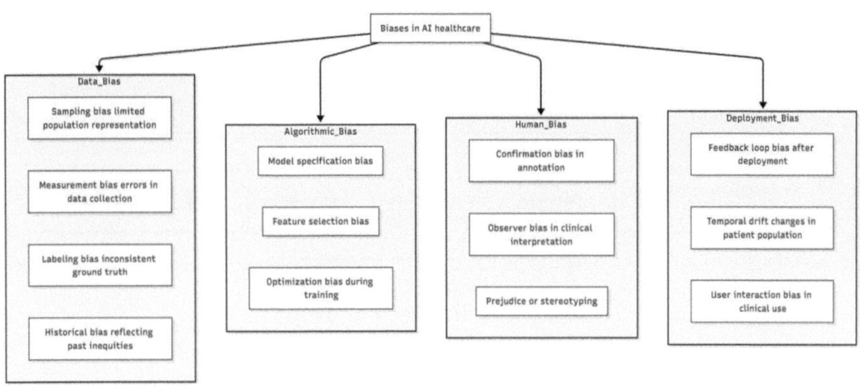


**Fig. 2.27** Biases in AI healthcare

these explanations. Confirmation bias can lead individuals to interpret results in a way that supports their preexisting beliefs. Depending on how they comprehend AI's explanations, people may continue to trust the AI, even when it is incorrect or unreliable [66]. Maintaining a realistic understanding of AI's capabilities and limitations enables medical staff to better navigate the ethical and legal challenges associated with AI in healthcare, while striking a balance between accountability and transparency.

## *2.7.1 Regulatory Frameworks and Compliance (FDA, EMA)*

A regulatory framework consists of a set of rules and regulations designed to ensure that individuals or organizations meet specific requirements. It also examines how these laws impact broader issues such as public health, safety, and justice. The **European Medicines Agency (EMA)** operates under European Union laws and employs an administrative procedure for pharmaceutical licensing. This allows a company to sell its products in any EU country by filing a single application. The primary focus of this system is to ensure the quality, safety, and effectiveness of the drugs sold in the EU [67]. In the United States, the Food and Drug Administration (FDA) supports federal laws, especially the Federal Food, Drug, and Cosmetic Act (FD&C Act). The FDA has several pathways for drug licensing, including the New Drug Application (NDA) and the Biologics License Application (BLA). These regulations aim to ensure that medications are effective and safe for use, thus protecting public health. Both the EMA and the FDA established regulations and standards that biotechnology companies must adhere to, which can vary widely. Therefore, it is crucial for these businesses to understand how each system operates, especially if they intend to offer products in both markets (Fig. 2.28).

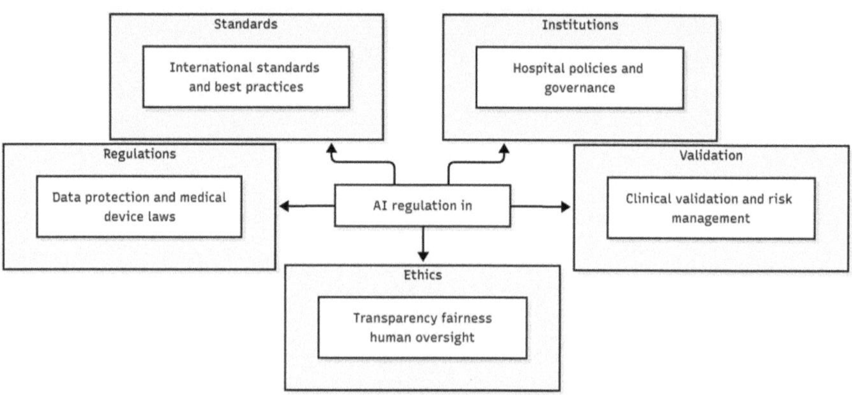


**Fig. 2.28** Regulatory framework

The healthcare sector is one of the most heavily regulated industries worldwide, with numerous laws that healthcare institutions and professionals must follow. These regulations are designed to ensure patient data security, provide high-quality care, and prevent fraud [68]. They cover a wide range of topics related to healthcare management and delivery.

Main Areas of Regulatory Compliance in Healthcare:

- Patient privacy and care: Legislation such as HIPAA protects patient data, medical records, and personal information. These laws also ensure that patients receive appropriate and equitable treatment. Violations can result in significant penalties.
- Worker safety: Regulations such as those enforced by OSHA protect healthcare workers from hazards, including exposure to radiation and infectious diseases. The aim is to preserve worker safety.
- Preventing fraud: Fraud is a significant issue in the healthcare sector, leading to substantial annual losses. Violating laws that prohibit fraud can have serious consequences, including the revocation of a medical license (Fig. 2.29).

The increased use of AI as a utility tool in various fields, including biomedicine, has raised skepticism regarding its fairness, bias, and legitimacy. This skepticism has led some scientists to dismiss AI as flawed and not yet ready for widespread application. Over the past decade, AI development has made significant strides, primarily driven by technological advancements in deep learning and neural networks. These advancements have resulted in sophisticated AI systems capable of performing critical tasks, such as interpreting medical literature at the level of expert physicians or devising recovery plans for patients with precision and efficiency. Previously, creating AI systems to assist pathologists, radiologists, and other specialists involved extensive engineering, design, and debugging of various algorithms, often taking years to complete [69]. However, recent advancements in artificial intelligence have enabled the training of deep learning algorithms using large sets of labeled or even

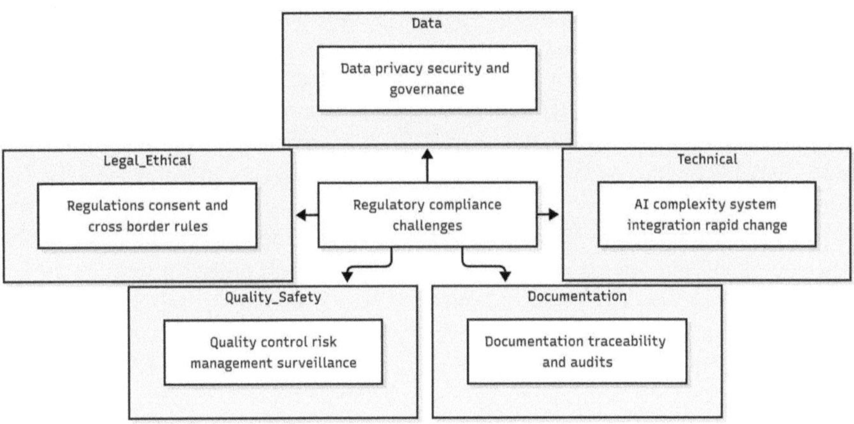


**Fig. 2.29** Challenges in regulatory compliance in biomedicine

unlabeled data. Deep learning approaches are highly adaptable to various tasks, allowing AI to transform and modernize healthcare by analyzing large datasets quickly and recognizing complex patterns. This capability could lead to a significant reduction in diagnostic errors, lower healthcare costs, and, most importantly, improved patient outcomes. However, this progress raises several ethical concerns regarding the reliance on computers for critical biomedicine decisions. If left unchecked, AI systems could perpetuate or even amplify existing biases, undermine the principles of fairness and transparency, and erode public trust in decision-making processes that affect the lives of many people.

Bias can be defined as a systematic or unfair disadvantage or favoritism towards certain individuals or groups, resulting in discrimination. It can manifest in various forms, including data bias, which originates from the datasets used to train AI systems. Recognizing the critical importance of addressing bias in AI, researchers and practitioners have developed several solutions to identify and mitigate bias, including [70]:

1. Data distribution: Ensuring that the data used for model training is evenly distributed, diverse, and valid. This involves actively seeking out and incorporating data from underrepresented groups, as well as applying techniques to mitigate the impact of sensitive attributes on model predictions.
2. Incorporating fairness Objectives: Integrating fairness objectives into AI models to promote equitable outcomes. This can be achieved by optimizing for demographic parity, equal opportunity, or other fairness measures to ensure that the model's decisions do not disadvantage or inconvenience any individual or group.
3. Bias auditing: Conducting comprehensive evaluations of generative AI systems to detect and address bias throughout their development. This includes testing for disparate impact, assessing model performance across different subgroups, and eliminating any potential sources of bias.

**Trustworthy AI systems** are essential in healthcare, where qualities such as transparency, explainability, and interpretability are crucial. These attributes ensure that models are safe, reliable, and fair across diverse patient populations. For healthcare providers to trust AI-generated insights, they must be confident that these insights are derived from representative and credible data sourced from legitimate origins, encompassing relevant demographic and clinical variations. A study examined factors that influence how much patients trust AI. It found that younger individuals are more concerned with ease of use, accountability, risk, and social influence, whereas older individuals prioritize practicality and are less sensitive to issues of risk and data privacy. This suggests that the direction of AI development should consider the expectations of all patients and incorporate feedback and criticism throughout the development process [71]. For an AI model to be considered trustworthy, it must be well-trained on verified data vetted by field experts. In instances involving rare health conditions, it is critical to ensure accurate data acquisition, verification, fact-checking, and that the information provided by the AI is trustworthy, reliable, and complex yet understandable. AI systems vary significantly in their ability to produce trustworthy information within the healthcare sector. Symbolic AI and rule-based expert systems offer greater explainability but often struggle with the complexity of healthcare. On the other hand, artificial intelligence models utilizing machine learning, particularly deep learning architectures, provide superior predictive power but face interpretability challenges, which affect their transparency. Hybrid approaches that combine advanced reasoning with deep learning have garnered attention for their ability to strike a balance between transparency and predictive performance, effectively merging the best of both worlds. Identifying the most suitable AI category for specific clinical tasks is key to ensuring maximal trust and accountability in healthcare AI applications.

**Transparency** is a vital ethical aspect of AI development, as it ensures accountability and builds user trust while enabling advanced control and governance over the AI's functions and outputs. The inherent operation mode of many AI systems presents challenges in identifying and addressing potential biases, errors, and unintended consequences due to a lack of interpretability. If these transparency issues remain unresolved, they can lead to undesirable outcomes that erode trust, pose safety risks, and hinder the adoption of clinical practices [72]. Thus, transparency is critical for fostering public trust in AI services; stakeholders and developers must establish clear metrics to enhance AI interpretability. Moreover, a lack of transparency can prevent necessary adjustments to maintain the AI, with the risk that an AI may begin self-reinforcing its results, potentially leading to total system collapse. To improve the transparency of AI algorithms, researchers have developed **Explainable AI (XAI),** a field dedicated to increasing the interpretability and explainability of AI systems, thus making their functions and architecture more accessible and understandable. Key XAI approaches include [73]:

- Developing interpretable models focused on transparency, such as decision trees or rule-based systems, which provide clear explanations for their outputs.
- Generating post hoc explanations using saliency maps or feature importance analysis.
- Conducting feature contribution analysis through SHAP (Shapley Additive Explanations).
- Utilizing Local Interpretable Model-agnostic Explanations (LIME) to approximate the AI model locally with an interpretable model for explaining individual predictions.
- Documenting and creating instructions and datasheets that provide clear information about an AI system's purpose, data sources, limitations, and potential biases.

While the development of Explainable AI represents a significant advance, challenges remain in achieving complete transparency. Simplifying complex models to enhance interpretability can sometimes lead to a loss of accuracy, resulting in oversimplified or misleading explanations. Additionally, the technical complexity of making certain AI models more interpretable can be daunting and requires considerable research and engineering effort [74]. Addressing these challenges necessitates collaboration among AI developers, domain experts, policymakers, and others to establish best practices, develop new methodologies, and ensure that transparency is upheld.

## References


1. Rehman MHU, Pinaya HL, Nachev P, Teo JT, Ourselin S, Cardoso MJ. Federated learning for medical imaging radiology. Br J Radiol. 2023;96(1150):20220890.
2. Guan H, Yap PT, Bozoki A, Liu M. Federated learning for medical image analysis: a survey. Pattern Recogn. 2024;151:110424.
3. Xu Z, Li J, Yao Q, Li H, Zhao M, Zhou SK. Addressing fairness issues in deep learning-based medical image analysis: a systematic review. npj Digit Med. 2024;7(1):286.
4. Koutsoubis N, Waqas A, Yilmaz Y, Ramachandran RP, Schabath MB, Rasool G. Privacy-preserving federated learning and uncertainty quantification in medical imaging. Radiol Artif Intell. 2025;7(1):e240637.
5. Li M, Jiang Y, Zhang Y, Zhu H. Medical image analysis using deep learning algorithms. Front Public Health. 2023;11:1273253.
6. Zhang H, Qie Y. Applying deep learning to medical imaging: a review. Appl Sci. 2023;13(18):10521.
7. Albuquerque C, Henriques R, Castelli M. Deep learning-based object detection algorithms in medical imaging: systematic review. Heliyon. 2025;11(1):e15521.
8. Kebaili A, Lapuyade-Lahorgue J, Ruan S. Deep learning approaches for data augmentation in medical imaging: a review. J Imaging. 2023;9(4):81.
9. Raza A, Guzzo A, Ianni M, Lappano R, Zanolini A, Maggiolini M, Fortino G. Federated learning in radiomics: a comprehensive meta-survey on medical image analysis. Comput Methods Programs Biomed. 2025;239:108768.
10. Bharati S, Mondal M, Podder P, Prasath VB. Deep learning for medical image registration: a comprehensive review. arXiv. 2022:2204.11341.
11. Li J, Chen J, Tang Y, Wang C, Landman BA, Zhou SK. Transforming medical imaging with transformers? A comparative review of key properties, current progresses, and future perspectives. Med Image Anal. 2023;85:102762.
12. Atasever S. In silico drug discovery: a machine learning-driven systematic review. Med Chem Res. 2024;33(9):1465–90.
13. Perumal M, Srinivas M. FMed-diffusion federated learning on medical image diffusion. bioRxiv. 2025:2025-04.
14. Sistaninejhad B, Rasi H, Nayeri P. Deep learning for medical image analysis: a review. Comput Math Methods Med. 2023;2023:7091301.
15. De Pretis F, Kervinen M, Haatainen K, Tiihonen M, Tolppanen AM, Forsberg MM. Collection of data for detecting adverse drug events in smart hospitals via electronic health records, the GTT, sensors and IoMT. Drug Saf. 2022;45(10):1161–2.
16. Badnjević A, Pokvić LG, Deumić A, Bećirović LS. Post-market surveillance of medical devices: a review. Technol Health Care. 2022;30(6):1315–29.
17. Siddiqui Z, Shukla VK, Sharma N. A review of the new European medical device regulations. Pharm Regul Aff. 2024;
18. Ucar A, Karakose M, Kırımça N. Artificial intelligence for predictive maintenance applications: key components, trustworthiness, and future trends. Appl Sci. 2024;14(2):898.
19. Ross JS, Blount KL, Ritchie JD, Hodshon B, Krumholz HM. Post-market clinical research conducted by medical device manufacturers: a cross-sectional survey. Med Devices Auckl. 2015;8:241.
20. Tarricone R, Callea G, Ogorevc M, Prevolnik Rupel V. Improving the methods for the economic evaluation of medical devices. Health Econ. 2017;26:70–92.
21. Badnjevic A, Gurbeta L, Jimenez ER, Iadanza E. Testing of mechanical ventilators and infant incubators in healthcare institutions. Technol Health Care. 2017;25(2):237–50.
22. Gurbeta L, Dzemic Z, Bego T, Sejdic E, Badnjevic A. Testing of anesthesia machines and defibrillators in healthcare institutions. J Med Syst. 2017;41(9):1-0.
23. Badnjević A, Cifrek M, Magjarević R, Džemić Z, editors. Inspection of medical devices: for regulatory purposes. Singapore: Springer; 2018.

24. Lembo S, Barra P, Di Biasi L, Bouwmans T, Tortora G. AI4RDD: artificial intelligence and rare disease diagnosis—a proposal to improve the anamnesis process. Image Vis Comput. 2025;162:105658.
25. Khude H, Shende P. AI-driven clinical decision support systems: revolutionizing medication selection and personalized drug therapy. Adv Integr Med. 2025:100529.
26. Pankow A, et al. Medical expert knowledge meets AI to enhance symptom checkers: opportunities for early rare disease detection. J Med Internet Res. 2025;27:e70512.
27. World Health Organization. WHO global report on rare diseases 2023. Geneva: World Health Organization; 2023.
28. Noll R, Berger A, Facchinello C, Stratmann K, Schaaf J, Storf H. Enhancing diagnostic precision for rare diseases using case-based reasoning. J Am Med Inform Assoc. 2025:ocaf092.
29. Weatherhead AJ, Wisk DF. Managing partner. GlobalMed Physicians; 2025.
30. Wojtara M, Rana E, Rahman T, Khanna P, Singh H. Artificial intelligence in rare disease diagnosis and treatment. Clin Transl Sci. 2023;16(11):2106–11.
31. Ilić N, Sarajlija A. Artificial intelligence in the diagnosis of pediatric rare diseases: from real-world data toward a personalized medicine approach. J Pers Med. 2025;15(9):407.
32. Jumper J, Evans R, Pritzel A, Green T, Figurnov M, Ronneberger O, et al. Highly accurate protein structure prediction with AlphaFold. Nature. 2021;596(7873):583–9.
33. Zhang K, Zhou HY, Baptista-Hon DT, Gao Y, Liu X, Oermann E, et al. Concepts and applications of digital twins in healthcare and medicine. Patterns. 2024;5(8):100782.
34. Dhudum R, Ganeshpurkar A, Pawar A. Revolutionizing drug discovery: a comprehensive review of AI applications. Drugs Drug Candidates. 2024;3(1):148–71.
35. Cui J, Wang M, Lin C, Xu X, Zhang Z. Exploring machine learning strategies for single-cell transcriptomic analysis in wound healing. Burns Trauma. 2025;13:tkaf032.
36. Li VO, Han Y, Kaistha T, Zhang Q, Downey J, Gozes I, et al. DeepDrug as an expert guided and AI driven drug repurposing methodology for selecting the lead combination of drugs for Alzheimer's disease. Sci Rep. 2025;15(1):2093.
37. Gao W, Emaminejad S, Nyein HYY, Challa S, Chen K, Peck A, et al. Fully integrated wearable sensor arrays for multiplexed in situ perspiration analysis. Nature. 2016;529(7587):509–14.
38. Heikenfeld J, Jajack A, Rogers J, Gutruf P, Tian L, Pan T, et al. Wearable sensors: modalities, challenges, and prospects. Lab Chip. 2018;18(2):217–48.
39. Templeton JM, Poellabauer C, Schneider S, Rahimi M, Braimoh T, Tadamarry F, et al. Modernizing the staging of Parkinson disease using digital health technology. J Med Internet Res. 2025;27:e63105.
40. Marin Fermin T, Migliorini F, Papakostas E, Al-Khelaifi K, Maldonado DR, Hovsepian JM, et al. Concomitant glenohumeral injuries in patients with distal clavicle fractures undergoing arthroscopic-assisted surgery: a systematic review. J Orthop Surg Res. 2022;17(1):31.
41. Ding Y, Udompanyawit C, Zhang Y, He B. EEG-based brain-computer interface enables real-time robotic hand control at individual finger level. Nat Commun. 2025;16(1):1–20.
42. Khojastehnezhad MA, Youseflee P, Moradi A, Ebrahimzadeh MH, Jirofti N. Artificial intelligence and the state of the art of orthopedic surgery. Arch Bone Jt Surg. 2025;13(1):17–22.
43. Coser O, Tamantini C, Soda P, Zollo L. AI-based methodologies for exoskeleton-assisted rehabilitation of the lower limb: a review. Front Robot AI. 2024;11:1341580.
44. Lee MH, Tian MY, Kim MK. The effectiveness of overground robot exoskeleton gait training on gait outcomes, balance, and motor function in patients with stroke: a systematic review and meta-analysis of randomized controlled trials. Brain Sci. 2024;14(8):834.
45. Nicora G, Pe S, Santangelo G, Billeci L, Aprile IG, Germanotta M, et al. Systematic review of AI/ML applications in multi-domain robotic rehabilitation: trends, gaps, and future directions. J Neuroeng Rehabil. 2025;22(1):79.
46. van der Kooij H, van Asseldonk EH, Sartori M, Basla C, Esser A, Riener R. AI in therapeutic and assistive exoskeletons and exosuits: influences on performance and autonomy. Sci Robot. 2025;10(104):eadt7329.

47. Dohle E, Swanson E, Jovanovic L, Yusuf S, Thompson L, Horsfall HL, et al. Toward the clinical translation of implantable brain–computer interfaces for motor impairment: research trends and outcome measures. Adv Sci. 2025;12:e01912.
48. Liu XY, Wang WL, Liu M, Chen MY, Pereira T, Doda DY, et al. Recent applications of EEG-based brain-computer-interface in the medical field. Mil Med Res. 2025;12(1):14.
49. Liu Y, Wu X, Sang Y, Zhao C, Wang Y, Shi B, et al. Evolution of surgical robot systems enhanced by artificial intelligence: a review. Adv Intell Syst. 2024;6(5):2300268.
50. Ansari ZJ, Aher A, Thitame SN. Advancements in robotics and AI transforming surgery and rehabilitation. J Pharm Bioallied Sci. 2025;17(Suppl 1):S46–8.
51. Ueda D, Kakinuma T, Fujita S, Kamagata K, Fushimi Y, Ito R, et al. Fairness of artificial intelligence in healthcare: review and recommendations. Jpn J Radiol. 2024;42(1):3–15.
52. Chinta SV, Wang Z, Palikhe A, Zhang X, Kashif A, Smith MA, et al. AI-driven healthcare: a review on ensuring fairness and mitigating bias. arXiv. 2024:2407.19655.
53. Pecchia L, Maccaro A, Matarrese MA, Folkvord F, Fico G. Artificial intelligence, data protection and medical device regulations: squaring the circle with a historical perspective in Europe. Health Technol. 2024;14(4):663–70.
54. Aboy M, Minssen T, Vayena E. Navigating the EU AI act: implications for regulated digital medical products. NPJ Digit Med. 2024;7(1):237.
55. Hanna MG, Pantanowitz L, Jackson B, Palmer O, Visweswaran S, Pantanowitz J, et al. Ethical and bias considerations in artificial intelligence/machine learning. Mod Pathol. 2025;38(3):100686.
56. Santra S, Kukreja P, Saxena K, Gandhi S, Singh OV. Navigating regulatory and policy challenges for AI enabled combination devices. Front Med Technol. 2024;6:1473350.
57. Liu M, Ning Y, Teixayavong S, Liu X, Mertens M, Shang Y, et al. A scoping review and evidence gap analysis of clinical AI fairness. NPJ Digit Med. 2025;8(1):360.
58. Hasanzadeh F, Josephson CB, Waters G, Adedinsewo D, Azizi Z, White JA. Bias recognition and mitigation strategies in artificial intelligence healthcare applications. NPJ Digit Med. 2025;8(1):154.
59. Li J, Chen G, Li G, Xiao L, Jia R, Zhang K. Flexible brain electronic sensors advance wearable brain–computer interface. NPJ Biomed Innov. 2025;2(1):24.
60. Spinelli JC, Suleski BJ, Wright DE, Grow JL, Fagans GR, Buckley MJ, et al. Wearable microfluidic biosensors with haptic feedback for continuous monitoring of hydration biomarkers in workers. NPJ Digit Med. 2025;8(1):76.
61. Brankovic A, Cook D, Rahman J, Delaforce A, Li J, Magrabi F, et al. Clinician-informed XAI evaluation checklist with metrics (CLIX-M) for AI-powered clinical decision support systems. NPJ Digit Med. 2025;8(1):364.
62. Mastour H, Dehghani T, Moradi E, Eslami S. Explainable artificial intelligence for predicting medical students' performance in comprehensive assessments. Sci Rep. 2025;15(1):23752.
63. Azadi A, García-Peñalvo FJ. Optimizing clinical decision support system functionality by leveraging specific human-computer interaction elements: insights from a systematic review. JMIR Hum Factors. 2025;12(1):e69333.
64. Asciak L, Kyeremeh J, Luo X, Kazakidi A, Connolly P, Picard F, et al. Digital twin assisted surgery: concept, opportunities, and challenges. NPJ Digit Med. 2025;8(1):32.
65. Bronowicki K, Antoniuk-Majchrzak J, Malesza I, Możarowski W, Szymborska A, Pachuta B, et al. An attempt to evaluate the use of mixed reality in surgically treated pediatric oncology patients. NPJ Digit Med. 2025;8(1):262.
66. Mekki YM, Luijten G, Hagert E, Belkhair S, Varghese C, Qadir J, et al. Digital twins for the era of personalized surgery. NPJ Digit Med. 2025;8(1):283.
67. Oh N, Lim M, Kim B, Shin J, Park S, Rhu J, et al. AI-assisted intraoperative navigation for safe right liver mobilization in pure laparoscopic donor hepatectomy: an experimental multi-institutional validation study. Sci Rep. 2025;15(1):27935.

68. Botonis OK, Mendley J, Aalla S, Veit NC, Fanton M, Lee J, et al. Feasibility of snapshot testing using wearable sensors to detect cardiorespiratory illness (COVID infection in India). NPJ Digit Med. 2024;7(1):289.
69. van Steijn NJ, Pepplinkhuizen S, Postema PG, Knops RE, Winter MM. Ventricular arrhythmia detection with a wearable ring-type PPG sensor: a feasibility study. Heart Rhythm. 2025;22(5):789–98.
70. Xue Z, Gai Y, Wu Y, Liu Z, Li Z. Wearable mechanical and electrochemical sensors for real-time health monitoring. Commun Mater. 2024;5(1):211.
71. Vaghasiya JV, Mayorga-Martinez CC, Pumera M. Wearable sensors for telehealth based on emerging materials and nanoarchitectonics. NPJ Flexible Electron. 2023;7(1):26.
72. Ates HC, Nguyen PQ, Gonzalez-Macia L, Morales-Narváez E, Güder F, Collins JJ, et al. End-to-end design of wearable sensors. Nat Rev Mater. 2022;7(11):887–907.
73. Jing AB, Garg N, Zhang J, Brown JJ. AI solutions to the radiology workforce shortage. NPJ Health Syst. 2025;2(1):20.
74. Wenderott K, Krups J, Zaruchas F, Weigl M. Effects of artificial intelligence implementation on efficiency in medical imaging—a systematic literature review and meta-analysis. NPJ Digit Med. 2024;7(1):265.

# Chapter 3
# Review & Practice Problems

This chapter consolidates the theoretical and applied knowledge from the previous chapters through structured summaries and practical exercises. It begins with a concise review of key concepts, algorithms, and biomedical case studies to reinforce understanding and highlight connections between computational methods and healthcare applications (Sect. 3.1). A diverse set of practice problems and exercises follows, encouraging critical thinking and the development of practical skills for implementing intelligent systems in biomedical engineering and related fields (Sect. 3.2).

## 3.1 Summary of Key Concepts

| | |
|---|---|
| Accuracy | Ratio of correctly classified instances to total predictions. |
| Activation function | Non-linear transformation applied to a neuron's weighted input (e.g., sigmoid, tanh, ReLU). |
| Activation saturation | Region where activation derivatives are near zero, hindering gradient flow. |
| AdaBoost | Boosting that reweights misclassified samples at each iteration. |
| Adam optimizer | Combines momentum and adaptive learning rates for fast, stable convergence. |
| Adaptation | Weight update process bringing BMU and neighbors closer to the input vector. |
| Adaptive learning rate methods | Optimizers (AdaGrad, RMSProp, Adam) that adjust learning rates per parameter. |



A. Badnjević, L. Spahić, *Intelligent Systems in Biomedicine*, Series in BioEngineering, https://doi.org/10.1007/978-3-032-14785-1_3

| | |
|---|---|
| Adaptivity | Altering behavior in response to new data or context. |
| Adjusted Rand Index (ARI) | Similarity measure between clustering and ground truth corrected for chance. |
| Advanced supervised learning | Techniques that improve performance/reliability of labeled-learning models beyond basic algorithms. |
| Advanced unsupervised learning | Field focused on discovering hidden structures, reducing dimensionality, and organizing unlabeled data. |
| Advantages of DSS | Better decision quality, faster decision-making, improved productivity, and integrated data management. |
| AI Surgical Planning | Use of AI to analyze imaging and patient data to optimize surgical strategy. |
| AI-Driven CDSS | Clinical decision support systems enhanced by machine learning, deep learning, and natural language processing to analyze big data and provide real-time, personalized recommendations. |
| Algorithm aversion | Distrust of automated systems after observing errors. |
| Algorithmic Bias | Bias arising from model design or optimization objectives. |
| AlphaFold | AI system that predicts high-accuracy 3D protein structures to guide drug discovery. |
| Amperometric Biosensor | Electrochemical sensor measuring current generated by redox reactions at a fixed potential. |
| Analog Filter | Circuit using capacitors and inductors to filter continuous signals. |
| Anomaly Detection | Identification of unusual patterns in time-series data indicating potential risks or malfunctions. |
| Anomaly detection | Identifying rare/unusual instances that deviate from normal patterns. |
| Apriori | Bottom-up frequent itemset mining using support thresholds. |
| Area Under Curve (AUC) | Scalar measure of classifier separability. |
| Artificial General Intelligence (AGI) | Hypothetical AI with human-level reasoning and adaptability across tasks. |
| Artificial Intelligence (AI) | Computer systems capable of performing tasks that normally require human intelligence, such as reasoning, learning, perception, and decision-making. |

| | |
|---|---|
| Artificial Narrow Intelligence (ANI) | "Weak AI" designed for a specific task (e.g., chatbots, medical image analysis). |
| Artificial neural network (ANN) | Computing model inspired by biological neurons, consisting of layers of interconnected nodes. |
| Artificial Superintelligence (ASI) | Theoretical AI exceeding human intelligence in all domains. |
| Association rule learning | Discovers co-occurrence patterns in transactional data. |
| Attention mechanism | Module that weights input features by relevance, focusing computation on important elements. |
| Attention weights | Learned coefficients indicating importance of input tokens or features. |
| AUC-ROC | Area under the ROC curve measuring trade-off between sensitivity and specificity. |
| Augmented intelligence | Real-time AI assistance that supports rather than replaces clinician decision-making. |
| Augmented Reality (AR) Surgery | Overlay of digital anatomical information on the surgical field for enhanced visualization. |
| AUPRC | Area under the Precision–Recall curve; summary metric suited to imbalanced datasets. |
| Autoencoder | Neural network trained to encode input into a latent representation and reconstruct it. |
| Automation bias | Over-reliance on automated outputs. |
| Automation of Decisions | Delegation of time-sensitive decisions to AI systems for faster responses. |
| Autonomy | Operating with limited human intervention. |
| Backpropagation | Algorithm computing gradients efficiently by propagating errors backward through the network. |
| Bagging (bootstrap aggregating) | Parallel, homogeneous ensembling using bootstrap resampled datasets to train base learners. |
| Batch normalization (BN) | Normalizes layer activations within each mini-batch to stabilize and speed training. |
| Bayesian network | Graphical model representing probabilistic dependencies among variables. |
| Bayesian optimization | Builds a surrogate model of performance to choose the next promising configs. |
| Best Matching Unit (BMU) | Neuron with weight vector most similar to the current input vector. |
| Bias | Parameter that shifts the neuron's activation threshold. |

| | |
|---|---|
| Bias Auditing | Systematic testing of AI models to detect and correct unfair performance across subgroups. |
| Big Data Analytics | Techniques for processing massive healthcare datasets to identify patterns and generate predictive insights. |
| Binomial/Multinomial/ Ordinal logistic | Two-class, multi-class, or ordered-class logistic regression variants. |
| Biosensor | Analytical device combining a biological recognition element with a transducer to convert biochemical signals into measurable electrical outputs. |
| Bisector method | Defuzzification selecting a value dividing the output fuzzy set into two areas of equal weight. |
| Bland–Altman plot | Graphical method to assess agreement between AI measurements and human experts. |
| Boosting | Sequential ensembling where each new learner emphasizes previous mistakes; predictions are weighted. |
| Borderline-SMOTE/MWMOTE | SMOTE variants focusing on hard-to-learn or boundary samples. |
| Brain-Computer Interface (BCI) | System enabling direct communication between the brain and external devices using neural signals. |
| Calibrated trust | Balanced reliance on AI outputs according to their reliability. |
| Calinski–Harabasz index | Ratio of between-cluster to within-cluster dispersion. |
| Causal Graph/Directed Acyclic Graph (DAG) | Visual representation of variable relationships used to identify confounding factors and estimate causal effects. |
| Causal Inference | Statistical methods (e.g., instrumental variable analysis, causal graphs) to identify cause–effect relationships in medical data. |
| CE mark | European conformity mark indicating a product meets health, safety, and environmental standards. |
| Cell-Based Biosensor | Sensor employing living cells to detect environmental or physiological changes. |
| Centroid method | Defuzzification technique computing the center of gravity of the output fuzzy set. |

|  |  |
|---|---|
| Class imbalance | One class far outnumbers another, biasing learners toward the majority. |
| Classification | AI-based assignment of labels (e.g., disease presence) to images or image regions. |
| Clinical Decision Support Framework | Structured approach to designing CDSS with principles such as accuracy, transparency, user-centered design, privacy, workflow integration, scalability, regulatory compliance, and bias mitigation. |
| Clinical Decision Support System (CDSS) | Computer-based tool that aids healthcare professionals in making evidence-based clinical decisions by providing timely, patient-specific recommendations. |
| Closed-Loop Rehabilitation | Adaptive therapy system that senses patient performance and adjusts assistance in real time. |
| Cloud PACS | Cloud-hosted PACS enabling remote access and telemedicine integration. |
| ClusterGAN | GAN with discrete latent variables enabling joint clustering and sample generation. |
| Clustering | Machine learning method that groups similar patient data points to reveal hidden patterns. |
| colsample_bytree | Fraction of features sampled per tree. |
| Competition | Stage where neurons compete to match the input and identify the BMU. |
| Computed tomography (CT) | Imaging method using X-ray slices to create detailed cross-sectional 3D images. |
| Computerized Maintenance Management System (CMMS) | Software for tracking and scheduling device maintenance. |
| Confidence | Conditional probability of the rule's consequent given its antecedent. |
| Confusion matrix | Table of true positives, false positives, true negatives, and false negatives. |
| Continual Learning | AI capability to update a digital twin by incorporating new data without retraining from scratch. |
| Continuous Glucose Monitor (CGM) | Wearable sensor providing real-time glucose levels for diabetes management. |
| Contractive autoencoder (CAE) | Autoencoder with a penalty on encoder gradients to enforce robustness. |

| | |
|---|---|
| Convergence | Point in training where loss stabilizes and parameters stop changing significantly. |
| Convolution | Operation applying shared kernels to extract local features. |
| Convolutional Neural Network (CNN) | Feedforward network using convolutional filters and pooling to process grid-structured data like images. |
| Cooperation | Stage where the BMU's neighbors are updated to preserve topological structure. |
| Core/border/noise points | DBSCAN categories for dense, fringe, and outlier points. |
| Cost-sensitive learning | Assigns higher penalties to costly errors (e.g., false negatives). |
| Co-Training Deep Autoencoder (Co-AE) | Multi-view autoencoder with co-training regularizer enforcing cross-view agreement. |
| Credit risk assessment (fuzzy) | Financial decision system handling vague or incomplete financial indicators. |
| Crisp logic | Classical binary logic where statements are strictly true or false. |
| Cross-entropy loss | Measures dissimilarity between predicted probability distribution and true labels for classification. |
| Cross-validation | Repeated train/validation splits for reliable performance estimates. |
| Cuffless Blood Pressure Monitor | Wearable sensor estimating blood pressure without inflatable cuffs. |
| Curse of dimensionality | Loss of meaningful distance measures in high-dimensional spaces. |
| da Vinci Surgical System | Widely used robotic platform providing 3D visualization and tremor-filtered instrument control. |
| Data Bias | Bias introduced when training datasets underrepresent or misrepresent certain populations. |
| Data Compatibility Challenge | Difficulty of integrating multi-omics datasets from different platforms and formats. |
| Davies–Bouldin index | Evaluates cluster compactness and separation. |
| DBSCAN | Density-based clustering that finds arbitrary-shape clusters and labels noise. |
| Decision Support System (DSS) | Broader system that collects and analyzes data to assist in decision-making for complex problems. |

| | |
|---|---|
| Decision Tree/Random Forest/XGBoost | ML algorithms commonly used for predictive maintenance of medical devices. |
| Decoder | Part of an autoencoder that reconstructs input from latent variables. |
| Deep Adaptive Clustering (DAC) | Treats clustering as pairwise similarity classification using a CNN. |
| Deep Belief Network (DBN) | Layered generative model trained by stacking Restricted Boltzmann Machines. |
| Deep Boltzmann Machine (DBM) | Generative model with stacked stochastic layers for hierarchical feature learning. |
| Deep Canonical Correlation Analysis (DCCA) | Learns nonlinear correlated representations across multiple views. |
| Deep clustering | Joint representation learning and clustering using deep neural networks. |
| Deep Clustering Network (DCN) | Jointly optimizes k-means clustering and autoencoder reconstruction. |
| Deep Constrained Clustering Network (DCCN) | Uses must-link and cannot-link constraints during clustering. |
| Deep Embedded Clustering (DEC) | Autoencoder-based model optimizing KL divergence between soft cluster assignments and a target distribution. |
| Deep learning (DL) | ML subfield using multi-layer neural networks to learn hierarchical data representations. |
| Deep Multi-View Spectral Clustering (DMVSC) | Combines deep feature learning with spectral graph clustering across views. |
| Deep Semi-Supervised Clustering (DSSC) | Autoencoder with clustering and small labeled guidance. |
| DeepCluster | Alternates CNN feature extraction with k-means clustering to create pseudo-labels. |
| DeepWalk | Random walk–based node embedding using a Skip-Gram model. |
| Defuzzification | Conversion of fuzzy outputs into a single crisp value (e.g., centroid method). |
| Demographic Parity/Equal Opportunity | Fairness metrics used to evaluate and mitigate bias in AI systems. |
| Dendrogram | Tree diagram visualizing hierarchical clustering merges/splits. |
| Denoising autoencoder (DAE) | Autoencoder trained to reconstruct clean inputs from corrupted versions. |

| | |
|---|---|
| Design steps of fuzzy controller | Identify variables, define membership functions, create rule base, implement inference/defuzzification, test and tune. |
| Dice coefficient/Jaccard index | Overlap metrics used for evaluating image segmentation accuracy. |
| DICOM | Standard for formatting and transmitting medical imaging data. |
| Digital Filter | Algorithm that processes digitized signals to remove noise or extract desired frequency components. |
| Digital pathology | Whole-slide imaging of tissue samples for computer-assisted analysis. |
| Digital twin | Virtual representation of a medical device reflecting real-time operational data. |
| Dilated convolution | Convolution using spaced kernel elements to enlarge receptive field without increasing parameters. |
| Dimensionality reduction | Compresses features while retaining salient structure. |
| DISERDIS | User-centered CDSS prototype developed within the MIRACUM network to assist in rare disease diagnosis by comparing patient data and identifying similar cases. |
| Distance metric | Rule for measuring closeness (e.g., Euclidean, Manhattan, Minkowski). |
| DNA/Nucleic Acid Biosensor | Sensor detecting specific DNA or RNA sequences through hybridization. |
| Domain adaptation | Adjusting models/features to reduce distribution shift between source and target. |
| Double Machine Learning/Causal Forests | Machine learning methods that improve causal effect estimation in complex, high-dimensional datasets. |
| Dropout | Randomly disables neurons during training to force robust feature learning. |
| Drug Repurposing (Drug Repositioning) | Identifying new therapeutic uses for existing drugs to reduce development time and cost. |
| Dual-Slope ADC | ADC with slow conversion but high accuracy, ideal for narrow-band biosignals. |
| Dynamic Learning | Continuous model updating as new patient data becomes available to improve predictions. |
| Dynamic Time Warping (DTW) | Algorithm for aligning time-series data to detect temporal anomalies. |

| | |
|---|---|
| Early stopping | Halts training when validation performance stops improving. |
| Edge Computing | Processing of data near the source (on-device) to reduce latency and power consumption. |
| EFB | Exclusive Feature Bundling; combines sparse, mutually exclusive features to reduce dimensionality. |
| Elbow method | Heuristic to choose k by locating the WCSS "elbow" point. |
| Electrochemical Biosensor | Sensor that measures electrical changes caused by biochemical reactions at electrodes. |
| Electronic Health Record (EHR) | Digital patient record integrating imaging and clinical data. |
| EMA | European Medicines Agency regulating the safety and efficacy of pharmaceuticals and AI-enabled devices in the EU. |
| Encoder | Part of an autoencoder that maps input data to latent variables. |
| Ensemble learning | Combining multiple models to produce a stronger predictor than any single model. |
| Parallel ensemble | Models trained independently and combined (e.g., by averaging or voting) to reduce variance. |
| Entropy | Measure of impurity or disorder in a dataset. |
| Enzyme-Based Biosensor | Biosensor using enzymes to catalyze reactions for detecting metabolites like glucose or lactate. |
| Epoch | One full pass through the entire training dataset. |
| eps (ε) | Neighborhood radius parameter in DBSCAN. |
| Ethical AI | AI designed to meet principles of fairness, accountability, and non-maleficence in healthcare. |
| Ethical Challenges in AI Deployment | Risks of discrimination, loss of privacy, overreliance on algorithms, and erosion of patient trust if AI systems are biased or poorly explained. |
| Exoskeleton | Wearable robotic frame assisting or augmenting human movement for rehabilitation or industrial use. |
| Explainability–Accuracy Trade-off | Tension where making AI more interpretable can reduce predictive performance. |

|  |  |
|---|---|
| Explainable AI (XAI) | Methods (e.g., LIME, SHAP) that clarify how AI models reach their predictions to increase clinical trust. |
| Exploding gradient | Gradients grow uncontrollably during backpropagation, causing unstable weight updates. |
| External evaluation | Cluster assessment comparing results to known labels. |
| Extreme Edge Computing | Computation performed directly on the sensor device without cloud transmission. |
| F1 score | Harmonic mean of precision and recall. |
| Fairness/accountability/transparency (FAT) | Practices to identify and reduce bias and explain model outcomes. |
| FDA | U.S. agency overseeing drug/device approval and AI-based medical products. |
| Feature engineering | Manual creation of input variables to improve ML model performance. |
| Feature extraction | Automatic learning of informative representations from raw input data. |
| Feature-based transfer | Learning shared representations that align source and target distributions. |
| Federated learning | Decentralized training where local models share updates but not raw patient data. |
| Feedforward neural network | Network where signals move only from input to output with no cycles. |
| Field Effect Transistor (FET) | Three-terminal device controlling current flow to amplify biosensor signals. |
| Fine-tuning | Continuing training of a pretrained model on a target dataset. |
| First-order logic | Formal language of predicates and quantifiers for explicit reasoning. |
| Flash ADC | High-speed analog-to-digital converter using parallel comparators for rapid conversion. |
| Foundation model | Large pretrained deep learning model adaptable to various medical imaging tasks. |
| FP-Growth | Frequent pattern mining via compressed FP-tree without exhaustive candidate generation. |
| FP-tree | Compact prefix tree storing itemset counts for FP-Growth. |
| Frames/schemas | Structured templates describing stereotypical situations or objects. |
| Fully connected layer | Layer where each neuron connects to all neurons in the previous layer. |

| | |
|---|---|
| Fuzzification | Conversion of crisp inputs into fuzzy values using membership functions. |
| Fuzzy AND (conjunction) | Minimum of the membership degrees of input conditions. |
| Fuzzy automotive control | Use of fuzzy logic for automatic transmission, braking, or cruise control. |
| Fuzzy control system | Controller using fuzzy logic to handle uncertain or nonlinear processes. |
| Fuzzy HVAC control | Fuzzy controller regulating heating, ventilation, and air conditioning with smooth adjustments. |
| Fuzzy Inference System (FIS) | Framework using fuzzification, rule base, inference, and defuzzification to map inputs to outputs. |
| Fuzzy logic | Logic with graded truth values (0–1) for imprecise concepts. |
| Fuzzy NOT (negation) | Complement of membership degree, calculated as $1 - \mu(x)$. |
| Fuzzy OR (disjunction) | Maximum of the membership degrees of input conditions. |
| Fuzzy rule | IF–THEN statement describing relations between fuzzy inputs and outputs. |
| Fuzzy set | Set where each element has a membership degree between 0 and 1. |
| Gamma (min_split_loss) | Minimum loss reduction required to split a node (XGBoost). |
| Gated Recurrent Unit (GRU) | Simplified LSTM with fewer gates but similar ability to model sequences. |
| Gaussian membership function | Smooth bell-shaped MF defined by center c and width $\sigma$. |
| Gaussian Mixture Model (GMM) | Probabilistic soft clustering via a mixture of Gaussians (EM algorithm). |
| GDPR | EU regulation requiring consent, limited retention, and strict protection of sensitive health data. |
| Generalization | Ability of a trained model to perform well on unseen data. |
| Generative Adversarial Network (GAN) | AI model that generates new compounds by competing a generator against a discriminator. |
| Genetic Algorithm (GA) | Optimization method inspired by natural selection, iteratively refining solutions to find the best outcome. |
| Genomics | Study of the complete set of an individual's genes and DNA sequences. |

|  |  |
|---|---|
| Gini impurity | Probability of incorrect classification of a randomly chosen element. |
| Goal-directed behavior | Selecting actions aligned with specified or learned objectives. |
| GOSS | Gradient-based One-Side Sampling; keeps large-gradient samples to speed training. |
| Grad-CAM | Visualization technique highlighting image regions that influence CNN predictions. |
| Gradient boosting | Boosting that fits new learners to the residual errors of the current ensemble. |
| Gradient clipping | Limits gradient magnitude to avoid exploding updates. |
| Gradient descent | Optimization method that updates parameters in the direction of the negative loss gradient. |
| Graph Neural Network (GNN) | AI model that analyzes relationships in knowledge graphs for drug-target prediction. |
| GraRep | Graph embedding capturing higher-order proximities through multi-step relationships. |
| Grid search | Exhaustive evaluation over a parameter grid. |
| Heterogeneous ensemble | Base learners use different algorithms. |
| Hexagonal topology | SOM layout where each neuron connects to six neighbors for smoother mapping. |
| Hidden layer | Intermediate layer transforming inputs before final output. |
| Hierarchical clustering | Builds nested clusters via agglomerative or divisive strategies. |
| HIPAA | U.S. law protecting PHI by setting national privacy and security standards for health data. |
| HITECH Act | U.S. law strengthening HIPAA with stricter data breach and security requirements. |
| HL7 | Standard for exchanging clinical and administrative healthcare information. |
| Homogeneous ensemble | All base learners use the same algorithm. |
| HOPE | Graph embedding allowing flexible similarity measures and asymmetric relations. |
| Human-centered AI | Design approach prioritizing human needs, control, and oversight. |
| Hybrid AI Systems | Combination of symbolic reasoning (rules) and deep learning to balance transparency with predictive power. |
| Hybrid Self-Training Clustering (HSTC) | Iteratively refines pseudo-labels with a small set of true labels. |

| | |
|---|---|
| Hyperparameter tuning | Searching parameter settings that govern learning (not learned from data). |
| ID3/C4.5/CART | Classic decision tree algorithms differing in split criteria and pruning methods. |
| Image acquisition | Process of capturing physical visual information and converting it into digital data. |
| Immuno Biosensor | Sensor using antibodies to recognize antigens or disease biomarkers. |
| Impedimetric Biosensor | Sensor measuring changes in electrical impedance due to biological interactions. |
| Inference engine | Mechanism applying fuzzy rules to fuzzified inputs to produce fuzzy outputs. |
| Information gain | Reduction in entropy achieved by a split in a decision tree. |
| Information Maximizing Self-Augmented Training (IMSAT) | Deep clustering maximizing mutual information while enforcing perturbation invariance. |
| In-Memory Computing | Performing computation within the memory element to minimize data transfer and power use. |
| Instance-based transfer | Reweighting/reusing source samples helpful to the target. |
| Instrumental Variable (IV) Analysis | Statistical method using external variables to infer causal relationships from observational data. |
| Intelligent Medical Maintenance Information System (IMMIS) | AI-enhanced CMMS with digital twins and predictive analytics. |
| Intelligent system | Computational entity that perceives its environment, reasons over data, learns from experience, and acts to achieve defined goals under uncertainty. |
| Internal covariate shift | Change in layer input distributions during training that slows convergence. |
| Internal evaluation | Cluster assessment using data-intrinsic metrics without labels. |
| Internal validation | Model testing on held-out data from the same distribution as training data. |
| Interpretability | Degree to which humans can understand a model's mechanisms or predictions. |
| Intraoperative Navigation | AI-guided real-time tracking of instruments relative to patient anatomy during surgery. |

| | |
|---|---|
| Invasive Neural Interface | Implanted electrodes providing high-resolution decoding of brain activity. |
| IoT fuzzy control | Fuzzy logic integrated with Internet of Things devices for adaptive control. |
| Isolation Forest | Randomly partitions data; anomalies isolate with shorter path lengths. |
| Isomap | Manifold learning preserving global geodesic distances. |
| Joint Multi-View Clustering (JMVC) | Learns a common latent space aligning clusters across views. |
| Joint Unsupervised Learning (JULE) | Recurrent process combining agglomerative clustering and representation refinement. |
| Kernel trick | Technique for creating non-linear decision boundaries by mapping data into higher-dimensional space implicitly. |
| k-fold CV | Splits data into k folds; each fold serves once as validation. |
| K-Means | Partitions data into k clusters by minimizing within-cluster variance. |
| k-Nearest Neighbors (k-NN) | Instance-based method predicting from the labels or values of the k closest training points. |
| Knowledge graph | Graph of entities and typed relationships for reasoning and retrieval. |
| Knowledge-Based CDSS | Decision support system that uses pre-defined medical rules and "if–then" logic encoded by experts. |
| Kohonen map | Another name for a Self-Organizing Map. |
| L1/L2 regularization (alpha/lambda) | Penalties on weights to control complexity. |
| L2 regularization (weight decay) | Adds squared weight penalties to the loss to discourage large parameters. |
| Laplacian Eigenmaps | Spectral method embedding nodes based on graph Laplacian eigenvectors. |
| Learning/Adaptation | Improving performance from data or feedback without explicit reprogramming. |
| Learning rate | Hyperparameter controlling step size in gradient descent updates. |
| Lift | Confidence relative to baseline prevalence; >1 indicates positive association. |
| Ligand-Based Virtual Screening (LBVS) | Screening based on similarity to known active compounds. |

| | |
|---|---|
| LightGBM | GBM using histogram-based, leaf-wise growth with GOSS and EFB for speed and memory efficiency. |
| LINE | Graph embedding preserving first- and second-order node proximities. |
| Linear/Polynomial/RBF/Sigmoid kernel | Common kernel functions for SVMs. |
| Linear regression | Model that fits a linear relationship between predictors and a continuous target. |
| Linguistic variable | Variable whose values are words or phrases (e.g., "cold," "warm," "hot"). |
| Local Outlier Factor (LOF) | Scores points by comparing local density to neighbors' densities. |
| Locally Linear Embedding (LLE) | Preserves local neighborhood relationships in a lower-dimensional embedding. |
| Logistic regression | Probabilistic classifier using the sigmoid function to model class probabilities. |
| Long Short-Term Memory (LSTM) | RNN variant with gating mechanisms that preserve long-term dependencies. |
| Loss function | Objective measuring discrepancy between predictions and targets (e.g., MSE, cross-entropy). |
| Machine Learning (ML) | AI method that trains algorithms to learn patterns and make predictions from data without explicit programming. |
| Magnetic resonance imaging (MRI) | Imaging technique using magnetic fields and radio waves to visualize soft tissues without ionizing radiation. |
| Mahalanobis distance | Distance scaled by covariance; flags points far from the multivariate mean. |
| Majority voting | Final class is the most common prediction among base models. |
| Mamdani inference system | FIS using fuzzy sets for outputs with defuzzification for crisp results. |
| Manifold learning | Techniques uncovering low-dimensional manifolds underlying high-dimensional data. |
| Margin | Distance between the separating hyperplane and the nearest data points of each class. |
| Matrix factorization | Graph embedding by decomposing similarity or adjacency matrices. |
| Max normalization | Scaling each feature by its maximum value to range [0,1]. |
| Max_depth | Maximum tree depth controlling model complexity. |

|  |  |
|---|---|
| Maximum membership principle | Defuzzification taking the value with the highest membership degree. |
| Mean Shift | Shifts candidate centers toward density peaks; finds modes without preset k. |
| Mean Squared Error (MSE) | Average of squared residuals; a regression loss metric. |
| Medical device (MD) | Instrument or apparatus used in healthcare for diagnosis, treatment, or monitoring of patients. |
| Medical fuzzy diagnosis | FIS applied to patient data with uncertain or incomplete information. |
| Membership function (MF) | Function $\mu(x)$ that assigns each element a membership value in a fuzzy set. |
| Message passing | Process in GNNs where nodes aggregate neighbor information to update their representations. |
| min_child_weight | Minimum sum of instance weights in a child node (XGBoost). |
| Mini-batch | Subset of training data used in one gradient descent update. |
| MinPts | Minimum points to form a dense region in DBSCAN. |
| Model Calibration | Adjustment of AI predictions to maintain alignment between the digital twin and real patient data. |
| Momentum | Gradient descent enhancement that smooths updates using a moving average of past gradients. |
| Motion Analysis | Quantitative assessment of patient movement patterns using sensors and AI. |
| Multimodal AI | Models combining multiple data types (imaging, genomics, clinical records) for integrated diagnostics. |
| Multimodal interaction | Interfaces combining text, speech, vision, or other input modes. |
| Multi-omics Integration | Combining genomic, transcriptomic, proteomic, and other biological data for comprehensive disease understanding. |
| Multiple instance learning (MIL) | Technique aggregating patch-level predictions to obtain slide-level decisions. |
| Multi-view deep clustering | Clustering that fuses representations from multiple data modalities. |
| n_estimators | Number of trees/boosting rounds in an ensemble. |

|  |  |
|---|---|
| Negative transfer | Transferred knowledge that harms target performance. |
| Nested CV | Outer CV for evaluation, inner CV for hyperparameter tuning. |
| Neural network (NN) | Computational model of interconnected neurons that learns functions by adjusting weighted connections. |
| Neural representation | Knowledge encoded implicitly in the parameters of a neural model. |
| Neuron (artificial) | Basic computing unit that sums weighted inputs, adds a bias, and applies an activation function. |
| Neuron grid | Lattice of units in a SOM where each neuron has an associated weight vector. |
| Neuroplasticity | Brain's ability to reorganize neural connections during recovery, enhanced by repetitive robotic therapy. |
| Neuro-symbolic AI | Hybrid systems combining neural learning with symbolic reasoning. |
| node2vec | Extension of DeepWalk with biased random walks to balance structural roles and community detection. |
| Non-Invasive BCI | Interface using EEG signals to control devices without surgical implantation. |
| Non-Knowledge-Based CDSS | Decision support system that uses machine learning algorithms (e.g., genetic algorithms, artificial neural networks) to learn patterns from data without explicit rules. |
| Normalized Mutual Information (NMI) | Mutual information between predicted clusters and true classes normalized for scale. |
| One-class SVM | Learns a boundary around normal data to detect outliers. |
| One-to-one/one-to-many/many-to-one/many-to-many | RNN input–output configurations for different sequence tasks. |
| Ontology | Formal specification of concepts and relationships in a domain. |
| Optical Biosensor | Sensor using light absorption, fluorescence, or reflectance to detect physiological changes. |
| Organic Electrochemical Transistor (OECT) | High-transconductance transistor using electrolytes to amplify low-frequency signals. |

| | |
|---|---|
| Organic Thin-Film Transistor (OTFT) | Printable, flexible transistor for skin-mounted sensors. |
| Overfitting | Model fits training data noise instead of general patterns, reducing performance on unseen data. |
| Parameter-based transfer | Sharing or initializing with parameters from pretrained models. |
| Patch-based learning | Analysis of large pathology slides by dividing them into smaller image patches. |
| Patient Stratification | Grouping patients based on shared traits or predicted outcomes to guide targeted therapies. |
| PCA | Linear projection onto principal components maximizing variance. |
| Perception | Acquiring and preprocessing information from sensors or inputs. |
| Performance evaluation (MD) | Collection and analysis of data on a device's real-world performance. |
| Personalized Medicine | Medical approach tailoring treatments to individual patient characteristics using genetic, clinical, and environmental data. |
| Photoplethysmography (PPG) | Optical method for measuring heart rate and blood oxygen saturation based on changes in blood volume. |
| Picture Archiving and Communication System (PACS) | Digital system for storing, retrieving, and sharing medical images. |
| PID controller | Classical Proportional-Integral-Derivative controller requiring precise system models. |
| Pipeline ADC | High-speed, high-precision ADC suited for demanding applications but complex for wearables. |
| Pooling | Downsampling operation reducing spatial dimensions while retaining salient information. |
| Positron emission tomography (PET) | Imaging method measuring metabolic activity with radiotracers to produce 3D functional images. |
| Post-hoc explanation | Technique that explains predictions of a complex model without exposing internal mechanics. |
| Post-market surveillance (PMS) | Continuous monitoring of device performance and safety after market release. |

|  |  |
|---|---|
| Potentiometric Biosensor | Sensor detecting changes in electrical potential without requiring current flow. |
| Powered Exoskeleton | Motorized exosuit providing active assistance for walking or lifting tasks. |
| Precision | Proportion of predicted positives that are actually positive. |
| Precision–Recall (PR) curve | Plot of precision vs. recall, useful with class imbalance. |
| Predictive maintenance (PdM) | AI-based maintenance strategy predicting failures before they occur. |
| Predictive Modeling for Treatment Outcomes | AI method integrating genetic, imaging, and clinical data to forecast a patient's response to therapy. |
| Pre-market process | Design, testing, approval, and certification steps before a medical device reaches the market. |
| Preprocessing (medical imaging) | Procedures (noise reduction, normalization, resizing, contrast enhancement) to improve and standardize images before analysis. |
| Pretrained model | Neural network trained on a large dataset and reused as a starting point for a new problem. |
| Pretraining | Training a model on a large source task before adapting to a target task. |
| Probabilistic modeling | Representing uncertainty with probability distributions. |
| Prospective validation | Real-time model testing within clinical workflows using current patient data. |
| Protected Health Information (PHI) | Sensitive personal medical data, including diagnoses, genetic records, and treatment details. |
| Pruning | Reducing tree size to improve generalization and prevent overfitting. |
| Pulse Oximeter | Device measuring blood oxygen saturation using optical absorption. |
| Purity | Proportion of cluster assignments matching the majority true class. |
| Quality management system (QMS) | Framework ensuring ongoing compliance and performance of medical devices. |
| Radiomics | Extraction of quantitative features from medical images to support predictive modeling. |

| | |
|---|---|
| Random Forest | Bagging of decision trees with random feature subsampling at each split; uses voting/averaging. |
| Random initialization | Setting SOM weights to small random values at training start. |
| Random oversampling (ROS) | Duplicate or resample minority-class instances. |
| Random search | Randomly samples parameter combinations from ranges/distributions. |
| Random undersampling (RUS) | Remove majority-class instances. |
| Random walk embedding | Learning node vectors from co-occurrence in short random walks (e.g., DeepWalk, node2vec). |
| Rare Disease (RD) | Condition affecting fewer than 1.3 in 2000 people, often genetic and difficult to diagnose. |
| Real-Time Analysis | Rapid processing of live data streams to support instant decisions. |
| Real-Time Decision Making (RTDM) | Process of making immediate, data-driven choices as events unfold using real-time analytics, AI, and automation. |
| Reasoning/Decision-making | Transforming information into conclusions or actions using models or logic. |
| Recall (Sensitivity) | Proportion of actual positives correctly identified. |
| Receptive field | Region of the input that affects a neuron's activation. |
| Recognition Element | Biological component (enzyme, antibody, nucleic acid, or cell) that specifically binds to a target analyte in a biosensor. |
| Rectangular topology | SOM layout where each neuron connects to four neighbors. |
| Recurrent neural network (RNN) | Network with feedback connections allowing sequence modeling and memory of past inputs. |
| Regression | Predicting continuous numeric outcomes from input features. |
| Regularization | Techniques to prevent overfitting by constraining model complexity. |
| Regularization hyperparameter ($\lambda$) | Coefficient controlling strength of penalty in L1 or L2 regularization. |
| Regulatory Compliance | Adherence to laws protecting patients, such as HIPAA, OSHA, and fraud-prevention regulations. |

| | |
|---|---|
| Regulatory Framework | Set of rules ensuring AI safety, data security, and public health compliance. |
| Rehabilitation Robotics | Robots used to assist and enhance physical therapy and motor recovery. |
| Reinforcement learning | ML approach where an agent learns actions by receiving rewards or penalties from the environment. |
| Reinforcement learning (RL) | Learning action policies through trial and error to maximize cumulative reward. |
| Relational-based transfer | Transferring structured relations/rules (e.g., graphs). |
| Residual | Difference between observed and predicted value. |
| ResNet/DenseNet | Deep CNN architectures used for medical image classification. |
| Robotic Surgery | Minimally invasive surgery performed with robotic arms controlled by a surgeon for enhanced precision. |
| Robustness | Maintaining performance despite noise, errors, or incomplete data. |
| ROC curve | Plot of true positive rate vs. false positive rate across thresholds. |
| Rule base | Collection of fuzzy rules forming the knowledge base of a fuzzy system. |
| SAR ADC | Successive approximation ADC offering low energy use and adjustable resolution. |
| Scale and shift parameters ($\gamma$, $\beta$) | Learnable BN parameters that rescale and offset normalized activations. |
| Segmentation | AI-driven delineation of anatomical structures or regions of interest within an image. |
| Self-Organizing Map (SOM) | Neural network that maps high-dimensional data to a low-dimensional grid while preserving topology. |
| Self-Supervised Learning | Model generating its own labels from data to identify anomalies. |
| Semantic network | Graph structure with nodes as concepts and edges as relationships. |
| Semi-Supervised DEC (SS-DEC) | DEC with an added supervised classification branch. |
| Semi-supervised deep clustering | Clustering guided by a small amount of labeled data. |
| Semi-Supervised Variational Deep Embedding (SS-VaDE) | VaDE incorporating labeled supervision for Gaussian mixture components. |

| | |
|---|---|
| Sensitivity (recall) | Proportion of true positives correctly identified by the model. |
| Sequential ensemble | Models trained in sequence so later models focus on earlier errors to reduce bias. |
| Sequential modeling tasks | Problems requiring ordered input processing (e.g., language, time series). |
| Shallow embedding | Graph embedding where each node is directly mapped to a low-dimensional vector without deep transformations. |
| Sigma-Delta ADC | Highly accurate ADC using oversampling for precise signal conversion. |
| Sigmoid function | S-shaped function mapping real values to probabilities between 0 and 1 |
| Signal Conditioning | Amplification and filtering of weak sensor signals to improve quality before digitization. |
| Silhouette score | Measures how similar a point is to its cluster compared to other clusters. |
| Single-view deep clustering | Learning and clustering from one data modality. |
| Sleep and Activity Tracker | Device recording movement, posture, and sleep quality for long-term health assessment. |
| Smart Prosthetic | Artificial limb with integrated sensors and AI for precise control and feedback. |
| Smart traffic system | Fuzzy system optimizing traffic light timing based on congestion measures. |
| SMOTE | Synthetic Minority Over-sampling Technique that generates new minority samples. |
| SMOTEBoost/AdaCost | Boosting methods tailored to imbalanced data. |
| Softmax function | Converts logits into a probability distribution over output classes. |
| Sparse autoencoder | Autoencoder encouraging most latent units to remain inactive. |
| Specificity | Proportion of true negatives correctly identified by the model. |
| Spiking Neural Network (SNN) | Event-driven neural network that processes data asynchronously to reduce energy consumption. |
| Stacking (stacked generalization) | Parallel, heterogeneous ensembling that trains a meta-learner on base model predictions. |
| Stratified CV | k-fold that preserves class proportions. |

|  |  |
|---|---|
| Structure-Based Virtual Screening (SBVS) | Screening based on the 3D structure of the target protein. |
| Subsample | Fraction of rows used per tree/boosting step. |
| Sugeno inference system | FIS using linear functions or constants for outputs with weighted averaging. |
| Supervised Anomaly Detection | Model trained with labeled normal and abnormal data for classification. |
| Supervised learning | Learning a mapping from inputs to labeled outputs. |
| Support | Frequency of an itemset in the dataset. |
| Support Vector Machine (SVM) | Classifier that finds the hyperplane maximizing the margin between classes. |
| Symbolic AI | Rule- and logic-based approaches that manipulate explicit symbols and relations. |
| Systematic initialization | Using principal component directions or sample vectors to initialize weights. |
| Tele-Rehabilitation | Remote robotic therapy delivered through cloud-based platforms and wearable sensors. |
| Thin-Film Transistor (TFT) | Lightweight, flexible transistor used in wearable electronics for signal amplification. |
| Transcriptomics | Study of RNA transcripts to determine which genes are actively expressed. |
| Transducer | Component that converts the biochemical interaction into an electrical or optical signal. |
| Transfer learning | Reusing a model trained on one task or domain to improve learning on a related task. |
| Transformer | Sequence model that replaces recurrence with self-attention for parallel processing of inputs. |
| Transparency | Ability to understand and explain how an AI model makes predictions. |
| Transparent model | Model whose internal logic is directly readable (e.g., decision tree). |
| Trapezoidal membership function | MF with a flat top defined by four points $(a < b \leq c < d)$. |
| Triangular membership function | MF defined by three points $(a < b < c)$ forming a triangle shape. |
| Trustworthy AI | AI that is transparent, reliable, and safe across diverse patient populations. |
| t-SNE | Nonlinear method preserving local similarity probabilities for visualization. |

| | |
|---|---|
| Turing machine | Abstract computational model introduced by Alan Turing, foundational to computer science. |
| Turing test | Evaluation of a machine's ability to exhibit human-like intelligence based on indistinguishable responses. |
| UKCA/CCC/PSE/CSA marks | National conformity certifications for medical devices in the UK, China, Japan, and Canada. |
| Ultrasound imaging | Technique using high-frequency sound waves to create real-time images of internal organs or tissues. |
| UMAP | Fast manifold learner preserving both local and global data structure. |
| Uncertainty communication | Presenting probabilities or confidence levels for user decision-making. |
| Undercomplete autoencoder | Autoencoder with a smaller latent space forcing feature extraction. |
| U-Net | CNN architecture designed for biomedical image segmentation with skip connections. |
| Unit-length normalization | Scaling so that each vector has norm equal to one. |
| Unsupervised Anomaly Detection | Model detecting deviations without labeled anomalies. |
| Unsupervised learning | Discovering structure or groupings in unlabeled data. |
| User-Centered Design (UCD) | Development approach involving end-users throughout the design process to ensure usability and acceptance. |
| Vanishing gradient | Gradients shrink toward zero in deep networks, slowing or halting learning in early layers. |
| Variational Autoencoder (VAE) | Generative model for creating new molecular structures with desired properties. |
| Variational Deep Embedding (VaDE) | VAE with a Gaussian mixture prior for probabilistic clustering. |
| Virtual Screening | Computational technique to identify candidate compounds likely to bind to a biological target. |
| Vision Transformer (ViT) | Neural architecture processing images as sequences of patches for classification. |
| Voltammetric Biosensor | Sensor recording current as voltage is swept to track oxidation–reduction reactions. |

| | |
|---|---|
| Wearable Biomedical Signal | Physiological signal (e.g., ECG, EEG, EMG, PPG) acquired by body-mounted sensors for health monitoring. |
| Wearable Biosensor | Portable sensor integrated into clothing or directly attached to the body for continuous monitoring of physiological parameters. |
| Weather prediction with fuzzy logic | Forecasting using imprecise meteorological data. |
| Weight | Trainable parameter that scales the influence of an input feature. |
| Weight initialization | Strategy for setting starting weight values (e.g., Xavier, He) to promote stable training. |
| Weighted voting | Base model votes are weighted by reliability/performance. |
| WHO ethical principles for AI | Global guidelines emphasizing autonomy, transparency, accountability, and inclusiveness. |
| XGBoost | Optimized GBM with regularization, shrinkage, and efficient split finding for high scalability. |
| X-ray imaging | Diagnostic technique using electromagnetic radiation to create 2D images of internal structures. |
| Z-score normalization | Standardization method subtracting mean and dividing by standard deviation for each feature. |

## 3.2 Practice Problems and Exercises

### *3.2.1 Introduction to Intelligent Systems*

1. Which of the following best distinguishes an intelligent system from traditional rule-based software?
   (a) Use of binary arithmetic only
   (b) Ability to perceive, reason, learn, and adapt
   (c) Dependence on explicit if–then statements for all scenarios
   (d) Operation without any human oversight
2. The "learning layer" of an intelligent system primarily enables:
   (a) Data acquisition through sensors
   (b) Pre-processing of noisy input

(c) Performance improvement through experience
(d) Actuator control in physical environments

3. Which learning paradigm requires labeled data for training?

(a) Reinforcement learning
(b) Supervised learning
(c) Unsupervised learning
(d) Self-supervised learning

4. A customer-segmentation algorithm that groups shoppers by purchasing habits without predefined categories is an example of:

(a) Supervised learning
(b) Unsupervised learning
(c) Reinforcement learning
(d) Hybrid learning

5. In reinforcement learning, the term **reward** refers to:

(a) The training dataset size
(b) Feedback signal guiding behavior
(c) The set of labeled examples
(d) The number of decision layers

6. Which property describes an intelligent system that continues functioning when sensor data are noisy or incomplete?

(a) Autonomy
(b) Goal-directed behavior
(c) Robustness
(d) Interpretability

7. A system that integrates neural networks for perception and symbolic logic for reasoning is an example of:

(a) Purely symbolic AI
(b) Neuro-symbolic or hybrid AI
(c) Purely statistical AI
(d) Fuzzy control

8. Which of the following methods explicitly handles uncertainty in knowledge representation?

(a) Frames
(b) Semantic networks
(c) Bayesian networks
(d) First-order predicate logic

9. In human–AI interaction, **automation bias** means:

(a) People distrust automation after small errors
(b) People over-rely on automated outputs even when wrong

(c) Machines ignore human feedback
(d) Algorithms avoid making uncertain predictions

10. Which domain typically requires interpretability to take precedence over raw predictive accuracy?

(a) Online advertising
(b) Movie recommendation
(c) Criminal justice risk assessment
(d) Social media content ranking

11. An intelligent system is a computational entity capable of ________, ________, ________, and acting upon the environment to achieve goals, often under uncertainty.
12. In a neural network trained for image recognition, the distributed numerical weights that capture patterns are an example of ________ representation.
13. The process of transforming raw data (e.g., pixels or audio waveforms) into features suitable for learning is known as ________.
14. A hospital deploys an AI tool that highlights suspicious regions in radiological images but cannot explain why it flagged them.
    *Question:* Identify the key human–AI interaction challenge and suggest one method to address it.
15. A start-up wants to detect fraudulent transactions but has very few labeled fraud cases and abundant unlabeled transaction data.
    *Question:* Which learning strategy (or combination) is most suitable, and why?

### *3.2.2 Supervised Learning Techniques*

1. In supervised learning, the training data must contain:

(a) Only input features without outputs
(b) Human-labeled input–output pairs
(c) Randomly generated outputs
(d) Features extracted automatically without any labels

2. When predicting house prices from attributes such as location and size, the problem type is:

(a) Classification
(b) Clustering
(c) Regression
(d) Reinforcement

3. Which of the following best describes the goal of linear regression?

(a) Assign class probabilities to categorical outputs
(b) Find a straight line that minimizes the sum of squared residuals

(c) Cluster unlabeled data into groups
(d) Separate classes using a hyperplane

4. In the linear regression equation y = mx + by = mx + by = mx + b, the parameter **m** represents:

   (a) Intercept of the line
   (b) Error term
   (c) Slope/gradient
   (d) Dependent variable

5. Logistic regression is most appropriate when the output variable is:

   (a) Continuous
   (b) Categorical
   (c) Unsupervised
   (d) Ordinal only

6. The function that maps real values to a range between 0 and 1 in logistic regression is called:

   (a) Rectifier
   (b) Hyperbolic tangent
   (c) Sigmoid
   (d) Exponential

7. In k-Nearest Neighbors, the parameter **k** determines:

   (a) The learning rate of gradient descent
   (b) The number of nearest data points used for prediction
   (c) The number of hidden layers in the network
   (d) The maximum depth of the decision boundary

8. Which distance metric measures the straight-line distance between two points?

   (a) Manhattan distance
   (b) Euclidean distance
   (c) Minkowski distance
   (d) Cosine similarity

9. Decision tree algorithms typically choose a feature split based on:

   (a) Maximum learning rate
   (b) Minimum residuals only
   (c) Impurity reduction (e.g., information gain or Gini)
   (d) Random selection of features

10. Among the major decision tree algorithms, which one always performs binary splits?

   (a) ID3
   (b) C4.5

(c) CART
(d) Random Forest

11. Support Vector Machines classify data by:

    (a) Computing Euclidean clusters
    (b) Finding a hyperplane that maximizes the margin between classes
    (c) Building ensembles of decision stumps
    (d) Minimizing information gain

12. The **kernel trick** in SVMs allows the algorithm to:

    (a) Reduce the number of support vectors
    (b) Map inputs to a higher-dimensional space without explicit transformation
    (c) Eliminate the need for regularization
    (d) Prune decision boundaries after training

13. Which evaluation metric becomes unreliable when classes are highly imbalanced?

    (a) Accuracy
    (b) Precision
    (c) Recall
    (d) F1 Score

14. The F1 score is defined as the harmonic mean of:

    (a) Accuracy and recall
    (b) Precision and recall
    (c) Precision and specificity
    (d) True positives and true negatives

15. The ROC curve plots:

    (a) Precision versus recall
    (b) False positive rate versus true positive rate
    (c) Accuracy versus loss
    (d) Training error versus testing error

16. In k-NN, a high value of **k** tends to ________ the decision boundary, reducing the risk of overfitting.
17. The confusion matrix provides counts of ________, ________, ________, and ________ to summarize classification performance.
18. A bank is testing two credit-risk models on a dataset where only 2% of applicants default. Accuracy is high for both, but the minority class is critical. Which evaluation metrics and visualization would you recommend and why?
19. A healthcare team wants a model that clearly explains its reasoning for patient diagnoses but can also handle both categorical and continuous features without heavy preprocessing. Which supervised learning method would you suggest and what precautions should be taken to avoid overfitting?

### 3.2.3 *Advanced Supervised Learning*

1. Ensemble learning improves prediction accuracy primarily by:
   (a) Training a single model on the entire dataset
   (b) Combining multiple models to reduce variance or bias
   (c) Increasing the size of the training dataset only
   (d) Using only unsupervised base learners
2. In a parallel ensemble, base models are:
   (a) Trained sequentially where each model corrects previous errors
   (b) Trained independently on different data subsets
   (c) Always different algorithms by design
   (d) Combined using gradient descent
3. Bagging is best described as:
   (a) Sequential learning with residual correction
   (b) Bootstrap sampling with parallel training of homogeneous models
   (c) Weighted voting with heterogeneous base learners
   (d) Hyperparameter tuning of a single model
4. Which of the following is a sequential ensemble method?
   (a) Random Forest
   (b) Bagging
   (c) Boosting
   (d) Stacking
5. In weighted majority voting, the weights assigned to classifiers are typically based on:
   (a) Random initialization
   (b) Historical reliability or validation accuracy
   (c) The size of the training dataset
   (d) The number of features used by each classifier
6. Stacking differs from bagging because:
   (a) Base learners are identical in stacking
   (b) A meta-learner combines base predictions
   (c) Sampling is always without replacement
   (d) Sequential training is mandatory
7. Random Forest reduces overfitting mainly by:
   (a) Using a single deep decision tree
   (b) Random feature selection and bootstrap sampling for each tree
   (c) Applying gradient descent to residual errors
   (d) Early stopping during tree growth only

8. During Random Forest training, each tree:
   (a) Uses all available features at every split
   (b) Uses a random subset of features for each split
   (c) Always outputs the same prediction
   (d) Must be pruned to a fixed depth
9. Which boosting variant adjusts sample weights of misclassified instances at each iteration?
   (a) Gradient Boosting
   (b) AdaBoost
   (c) LightGBM
   (d) Bagging
10. Gradient Boosting differs from AdaBoost because it:
   (a) Works only with decision stumps
   (b) Uses residual errors to guide new learners
   (c) Requires all base learners to be heterogeneous
   (d) Does not allow regularization
11. Compared to XGBoost, LightGBM is generally:
   (a) Slower and less memory-efficient
   (b) Faster with histogram-based leaf-wise growth
   (c) Restricted to small datasets only
   (d) Less prone to overfitting on small datasets
12. Which hyperparameter controls the learning rate in gradient boosting algorithms?
   (a) max_depth
   (b) n_estimators
   (c) learning_rate
   (d) colsample_bytree
13. Bayesian optimization in hyperparameter tuning is preferred when:
   (a) The search space is tiny and exhaustive search is cheap
   (b) Evaluations are expensive and a probabilistic model can guide sampling
   (c) Hyperparameters are categorical only
   (d) Random search consistently outperforms it
14. Stratified cross-validation is important because it:
   (a) Guarantees the smallest training error
   (b) Preserves class proportions across folds
   (c) Always reduces computation time
   (d) Removes the need for a validation set

15. Which metric is most appropriate for evaluating models on highly imbalanced datasets?
    (a) Overall accuracy
    (b) F1 score or AUC
    (c) Mean squared error
    (d) R-squared
16. SMOTE is an example of:
    (a) Algorithm-level cost-sensitive learning
    (b) Synthetic minority oversampling at the data level
    (c) Meta-learning ensemble strategy
    (d) Sequential boosting of weak learners
17. Cost-sensitive learning addresses imbalance by:
    (a) Oversampling the minority class
    (b) Assigning higher misclassification penalties to the minority class
    (c) Reducing the feature space dimensionality
    (d) Ignoring the majority class
18. Which transfer learning strategy reuses model parameters from a pre-trained network?
    (a) Instance-based transfer
    (b) Feature-based transfer
    (c) Parameter-based transfer
    (d) Relational-based transfer
19. A key risk of transfer learning is:
    (a) Bootstrapping bias
    (b) Negative transfer when domains differ significantly
    (c) Excessive variance in bagging
    (d) Underfitting due to regularization
20. LightGBM achieves faster training primarily because it:
    (a) Uses cost-sensitive loss exclusively
    (b) Employs histogram-based algorithms and leaf-wise growth
    (c) Avoids any form of regularization
    (d) Uses a single deep decision tree with cross-validation
21. Combining predictions from multiple base learners to improve accuracy is known as ________ learning.
22. Random Forest combines the outputs of many ________ to produce a final prediction.
23. The process of stopping training early to avoid overfitting is called ________.
24. When the knowledge from a source task degrades performance on a target task, it is known as ________.

25. A fraud detection dataset contains 99% non-fraud transactions.
    (a) Which sampling or algorithmic methods would you recommend to improve minority class detection?
    (b) Which evaluation metrics should be prioritized and why?
26. A data scientist wants to speed up model training on a 100-million-row dataset with limited GPU resources.
    (a) Which gradient boosting algorithm would you suggest?
    (b) Name two hyperparameters that should be carefully tuned to prevent overfitting.
27. A hospital needs a predictive model for rare disease detection but has only a small labeled dataset.
    (a) Which supervised learning strategy could reuse knowledge from a large, related dataset?
    (b) Mention one advantage and one risk of using this strategy.

### *3.2.4 Unsupervised Learning Techniques*

1. Unsupervised learning is primarily used to:
    (a) Predict continuous target values using labeled data
    (b) Detect patterns or structures in unlabeled data
    (c) Minimize overfitting in regression models
    (d) Train meta-learners from ensemble outputs
2. Which of the following is **not** an unsupervised learning task?
    (a) Clustering
    (b) Dimensionality reduction
    (c) Anomaly detection
    (d) Logistic regression
3. In K-Means, the parameter **K** indicates:
    (a) Number of nearest neighbors
    (b) Number of clusters to form
    (c) Learning rate of the algorithm
    (d) Maximum distance threshold
4. The **elbow method** is commonly used to:
    (a) Determine cluster quality in DBSCAN
    (b) Choose the optimal number of clusters in K-Means
    (c) Reduce dimensions before applying PCA
    (d) Set the bandwidth for Mean Shift

5. DBSCAN requires which two key parameters?
   (a) Support and confidence
   (b) Learning rate and momentum
   (c) Eps (neighborhood radius) and MinPts (minimum points)
   (d) Silhouette score and variance
6. A major strength of DBSCAN compared to K-Means is that it:
   (a) Forces clusters to be spherical
   (b) Automatically detects noise and clusters of arbitrary shape
   (c) Requires the number of clusters as input
   (d) Uses hierarchical linkage for merging clusters
7. Principal Component Analysis (PCA) reduces dimensionality by:
   (a) Selecting features with the highest variance directions
   (b) Maximizing cluster separation in local neighborhoods
   (c) Randomly projecting data to lower dimensions
   (d) Iteratively merging similar clusters
8. t-SNE is preferred over PCA when the goal is to:
   (a) Capture global linear relationships
   (b) Preserve local neighborhood structure for visualization
   (c) Perform feature scaling
   (d) Estimate covariance matrices
9. Which anomaly detection technique measures distance while accounting for correlations among features?
   (a) Euclidean distance
   (b) Mahalanobis distance
   (c) Isolation Forest
   (d) Gaussian Mixture Model
10. Association rule learning evaluates rule strength using:
    (a) F1 score, recall, precision
    (b) Support, confidence, lift
    (c) Eps, MinPts, bandwidth
    (d) Accuracy, sensitivity, specificity
11. In hierarchical clustering, the tree-like diagram used to visualize nested clusters is called a ________.
12. Isolation Forest identifies anomalies by constructing random trees and measuring the ________ length of data points.
13. In PCA, the new axes that capture the maximum variance in the data are called ________ components.
14. A retail chain wants to discover items frequently purchased together to optimize product placement.

(a) Which unsupervised learning algorithm is appropriate?
(b) Which metric indicates whether two products occur together more often than by chance?

15. A network security team needs to detect rare intrusions in massive traffic logs without labeled attack data.

(a) Name one unsupervised anomaly detection method suitable for this task.
(b) Give one reason this method is advantageous for high-dimensional data.

### 3.2.5 *Advanced Unsupervised Learning*

1. Self-Organizing Maps (SOMs) primarily perform:

(a) Supervised classification of labeled data
(b) Dimensionality reduction while preserving topological relationships
(c) Gradient boosting for prediction tasks
(d) Reinforcement learning through reward maximization

2. In SOM training, the neuron whose weight vector is closest to the input is called the:

(a) Principal Component Unit
(b) Best Matching Unit (BMU)
(c) Random Walk Node
(d) Discriminant Vector

3. Which of the following is **not** a typical step in SOM learning?

(a) Competition
(b) Cooperation
(c) Adaptation
(d) Backpropagation of labeled errors

4. A key reason for normalizing input data before SOM training is to:

(a) Increase the number of neurons in the output layer
(b) Ensure all attributes contribute equally to distance calculations
(c) Speed up t-SNE convergence
(d) Guarantee rectangular topology

5. Which topology is often preferred in SOMs for better neighborhood interaction?

(a) Circular
(b) Rectangular
(c) Hexagonal
(d) Triangular

6. Autoencoders differ from traditional PCA because they:
   (a) Require labeled target outputs
   (b) Can learn non-linear feature representations
   (c) Only operate with covariance matrices
   (d) Always reduce data to one dimension
7. Which type of autoencoder introduces noise during training to learn robust features?
   (a) Sparse Autoencoder
   (b) Contractive Autoencoder
   (c) Denoising Autoencoder
   (d) Variational Autoencoder
8. Variational Autoencoders (VAEs) are distinguished by their ability to:
   (a) Generate new data samples by learning a latent probability distribution
   (b) Cluster data directly using k-means loss
   (c) Guarantee deterministic latent codes
   (d) Perform hierarchical agglomerative merging
9. Deep Embedded Clustering (DEC) jointly optimizes:
   (a) Euclidean distance and Mahalanobis distance
   (b) Reconstruction loss and KL divergence clustering objective
   (c) Silhouette score and purity index
   (d) Gradient boosting and random walks
10. Which manifold learning algorithm explicitly preserves **global** pairwise distances?
    (a) Locally Linear Embedding (LLE)
    (b) Laplacian Eigenmaps
    (c) Isomap
    (d) UMAP
11. The unsupervised neural network introduced by Teuvo Kohonen that maps high-dimensional inputs to a 2-D grid is called a ________.
12. In sparse autoencoders, a ________ divergence penalty is often added to enforce low average activation of hidden units.
13. The internal evaluation metric that measures how similar an object is to its own cluster compared to other clusters is the ________ score.
14. A pharmaceutical company needs to visualize chemical compound relationships in a high-dimensional space and group similar molecules without labels.
    (a) Name one suitable **manifold learning** method.
    (b) Give one reason this method is preferred over PCA for this task.
15. A researcher wants to combine limited labeled tumor data with a large pool of unlabeled images to improve clustering quality.

(a) Identify one **semi-supervised deep clustering** approach.
(b) Explain briefly how labeled samples guide the clustering process.

### *3.2.6 Perceptrons and Multilayer Perceptrons (MLPs)*

1. The McCulloch–Pitts neuron (1943) introduced which key concept?
   (a) Adaptive weight learning
   (b) Binary threshold logic for neurons
   (c) Backpropagation of error signals
   (d) Non-linear activation with gradient descent
2. Rosenblatt's perceptron differs from the McCulloch–Pitts model because it:
   (a) Uses a continuous activation function
   (b) Incorporates an adaptive learning rule to update weights
   (c) Requires labeled data to compute gradients
   (d) Works only with non-linear decision boundaries
3. The perceptron convergence theorem guarantees that the perceptron will find a solution if:
   (a) The learning rate is exactly 0.01
   (b) The data are linearly separable
   (c) Dropout is applied during training
   (d) The network has at least two hidden layers
4. Which problem famously demonstrated the limitation of single-layer perceptrons?
   (a) AND
   (b) OR
   (c) XOR
   (d) NAND
5. The bias term in a perceptron primarily allows the model to:
   (a) Increase the learning rate
   (b) Shift the decision boundary away from the origin
   (c) Reduce overfitting in deep networks
   (d) Eliminate the need for weight updates
6. In an MLP, hidden layers use differentiable activation functions such as:
   (a) Step function only
   (b) Sigmoid, tanh, or ReLU
   (c) Hard limiters with no gradient
   (d) Polynomial basis expansions

7. Without non-linear activation functions, a multilayer perceptron would:
   (a) Behave as a universal approximator
   (b) Collapse into an equivalent single linear transformation
   (c) Automatically avoid overfitting
   (d) Require no gradient computation
8. Backpropagation enables MLP training by:
   (a) Performing random weight initialization
   (b) Propagating error gradients backward using the chain rule
   (c) Increasing the number of hidden neurons automatically
   (d) Replacing gradient descent with closed-form solutions
9. The universal approximation theorem states that an MLP with one hidden layer can:
   (a) Only solve linearly separable problems
   (b) Approximate any continuous function on a compact domain
   (c) Require no regularization
   (d) Learn without training data
10. Which activation function became popular for deep MLPs due to its resistance to vanishing gradients?
    (a) Sigmoid
    (b) Hyperbolic tangent (tanh)
    (c) Rectified Linear Unit (ReLU)
    (d) Step function
11. The neuron in an MLP layer computes a weighted sum of inputs plus a ________ term before applying an activation function.
12. The algorithm that updates perceptron weights only when a classification error occurs is called the ________ learning rule.
13. In backpropagation, the ________ rule of calculus is used to compute gradients of the loss with respect to each weight.
14. A dataset is not linearly separable.
    (a) Why will a single-layer perceptron fail on this data?
    (b) How does adding hidden layers in an MLP overcome this limitation?
15. A researcher trains a deep MLP and observes slow convergence and unstable gradients.
    (a) Name one modern **weight-initialization** method that can help.
    (b) Mention one **regularization** technique to reduce overfitting.

### 3.2.7 Fundamentals of Neural Networks

1. Which statement best describes a neural network?
   (a) A network of binary logic gates that stores explicit rules
   (b) A collection of interconnected neurons that learn by adjusting connection weights
   (c) A symbolic reasoning engine using decision trees
   (d) A statistical model limited to linear regression
2. The function of an activation function in a neuron is to:
   (a) Eliminate the need for weight updates
   (b) Introduce nonlinearity into the model
   (c) Perform gradient clipping during backpropagation
   (d) Guarantee convergence without training
3. Which activation function is defined as ReLU(z) = max(0, z)?
   (a) Sigmoid
   (b) Hyperbolic tangent (tanh)
   (c) Rectified Linear Unit
   (d) Softmax
4. Feedforward neural networks differ from recurrent neural networks because they:
   (a) Process sequences using feedback links
   (b) Contain cycles for temporal memory
   (c) Pass information only from input to output without feedback
   (d) Require convolutional layers for training
5. The softmax function is primarily used in the output layer for:
   (a) Binary regression problems
   (b) Multi-class classification to produce class probabilities
   (c) Weight regularization
   (d) Preventing vanishing gradients
6. A vanilla RNN struggles with long-term sequence dependencies mainly due to:
   (a) Lack of nonlinearity in hidden layers
   (b) Vanishing or exploding gradients
   (c) Overuse of dropout
   (d) Absence of softmax activation
7. Which RNN variant was specifically designed to handle long-range dependencies by using gating mechanisms?
   (a) Vanilla RNN
   (b) Long Short-Term Memory (LSTM)
   (c) Perceptron

(d) Convolutional Neural Network

8. In gradient descent, the parameter update rule $w \leftarrow w—\eta\ \partial L/\partial w$, $\eta$ represents:

   (a) Momentum factor
   (b) Learning rate
   (c) Dropout rate
   (d) Regularization constant

9. Which optimizer combines momentum with adaptive learning rates using first and second moment estimates?

   (a) AdaGrad
   (b) RMSProp
   (c) Adam
   (d) Huber

10. Dropout reduces overfitting primarily by:

    (a) Clipping large gradients during training
    (b) Randomly deactivating neurons to prevent co-adaptation
    (c) Scaling inputs to zero mean and unit variance
    (d) Increasing the learning rate dynamically

11. L2 weight regularization adds a penalty proportional to the ________ of the weights to the loss function.
12. During backpropagation, gradients are efficiently computed by applying the ________ rule of calculus.
13. Batch Normalization normalizes activations within each ________ to reduce internal covariate shift.
14. A deep network shows extremely slow learning in early layers.

    (a) Identify the gradient-related issue causing this behavior.
    (b) Name one architectural or initialization strategy that mitigates it.

15. A model trained on a small dataset achieves near-perfect accuracy on training data but performs poorly on new data.

    (a) What problem is occurring?
    (b) Give one regularization technique that addresses it.

### *3.2.8 Advanced Network Classifications*

1. Which key property distinguishes a Convolutional Neural Network (CNN) from a fully connected network?

   (a) Use of backpropagation for weight updates
   (b) Local connections and weight sharing

(c) Reliance on recurrent feedback loops
(d) Exclusive use of ReLU activation

2. In a CNN, the purpose of **padding** is to:

(a) Reduce overfitting by randomly dropping neurons
(b) Increase the receptive field without losing border information
(c) Accelerate convergence by normalizing activations
(d) Control gradient clipping during training

3. What does **stride** control in a convolutional layer?

(a) Number of filters per layer
(b) Step size of the convolution kernel across the input
(c) Dropout probability
(d) Batch size during training

4. Which pooling method selects the maximum value within each region of a feature map?

(a) Average pooling
(b) Global pooling
(c) Max pooling
(d) Dilated pooling

5. A primary advantage of **dilated convolution** is that it:

(a) Reduces the number of feature maps
(b) Expands the receptive field without increasing kernel size
(c) Eliminates the need for activation functions
(d) Prevents vanishing gradients

6. Recurrent Neural Networks (RNNs) differ from CNNs because they:

(a) Require convolutional filters
(b) Employ feedback loops to process sequential data
(c) Cannot handle variable-length inputs
(d) Always use softmax outputs

7. Long Short-Term Memory (LSTM) networks address which key limitation of standard RNNs?

(a) Lack of local receptive fields
(b) Vanishing and exploding gradients in long sequences
(c) Excessive parameter sharing
(d) Inability to perform backpropagation

8. Transformers outperform RNNs on long sequences mainly because they:

(a) Use sequential time-step processing
(b) Apply multi-head self-attention in parallel
(c) Depend on gated recurrent units for memory

(d) Require no optimization algorithm

9. In a Transformer, the **multi-head attention** mechanism allows the model to:

   (a) Randomly drop attention weights during training
   (b) Attend to multiple representation subspaces simultaneously
   (c) Reduce training time by removing position encoding
   (d) Eliminate the need for feedforward layers

10. Graph Neural Networks (GNNs) are particularly suited for:

    (a) Images represented as pixel grids
    (b) Sequential time-series forecasting
    (c) Data represented as nodes and edges with relational structure
    (d) Purely tabular numerical datasets with no relationships

11. CNNs reduce parameters by using ________ connections and ________ of filter weights across spatial locations.
12. The gating components that control information flow in an LSTM are the input gate, output gate, and ________ gate.
13. In transfer learning, a model trained on a large source dataset is reused and ________ on a smaller target dataset to improve performance.
14. A hospital wants to detect rare protein-interaction patterns from molecular graphs of chemical compounds.

    (a) Which neural network type is most appropriate?
    (b) Give one reason this architecture fits the problem.

15. An NLP team must translate long documents efficiently but finds RNN training unstable.

    (a) Which architecture should they adopt to overcome sequential bottlenecks?
    (b) Name the key mechanism that enables this model to capture long-range dependencies.

### *3.2.9 Fuzzy Systems and Their Applications*

1. Fuzzy logic differs from classical (crisp) logic primarily because it:

   (a) Requires binary inputs and outputs
   (b) Allows degrees of truth between 0 and 1
   (c) Uses only probabilistic reasoning
   (d) Eliminates the need for inference rules

2. Who introduced the concept of fuzzy sets and fuzzy logic in 1965?

   (a) Ebrahim Mamdani
   (b) Marvin Minsky

(c) Lotfi A. Zadeh
(d) Takagi Sugeno

3. A **membership function** in a fuzzy system:

   (a) Stores the training dataset
   (b) Maps each input to a degree of membership in [0, 1]
   (c) Selects the defuzzification method
   (d) Converts fuzzy outputs to crisp values

4. In fuzzy logic, the **AND** operation is commonly implemented using which mathematical operation?

   (a) Maximum
   (b) Minimum
   (c) Average
   (d) Difference

5. Which membership function provides a smooth bell-shaped curve suitable for gradual transitions?

   (a) Triangular
   (b) Trapezoidal
   (c) Gaussian
   (d) Step

6. What is the purpose of **defuzzification** in a fuzzy inference system?

   (a) Transform crisp inputs into fuzzy values
   (b) Combine fuzzy sets into linguistic variables
   (c) Convert fuzzy outputs into a single crisp value
   (d) Generate the rule base automatically

7. The Mamdani inference method differs from the Sugeno method because Mamdani:

   (a) Uses linear functions as rule consequents
   (b) Employs fuzzy sets as outputs before defuzzification
   (c) Requires no defuzzification
   (d) Cannot handle multiple inputs

8. Compared with PID controllers, fuzzy controllers are advantageous because they:

   (a) Require exact mathematical models of the system
   (b) Handle nonlinearities and uncertainties more effectively
   (c) Depend solely on gradient descent optimization
   (d) Cannot be tuned once deployed

9. Which defuzzification method calculates the center of the aggregated output set?

   (a) Maximum membership principle
   (b) Bisector method

(c) Centroid (center of gravity) method
(d) Weighted average of inputs

10. A hybrid intelligent system that combines fuzzy logic with neural networks primarily aims to:

    (a) Eliminate fuzzy variables
    (b) Enable the system to learn and adapt from data
    (c) Remove the need for membership functions
    (d) Restrict rule generation to expert input only

11. In fuzzy logic, linguistic variables such as "cold," "warm," and "hot" are represented by ________ sets with associated ________ functions.
12. The process of converting precise numerical inputs into fuzzy values is called ________, whereas converting fuzzy outputs back to crisp values is known as ________.
13. A trapezoidal membership function is defined by ________ parameters and contains a flat region of full membership.
14. A smart HVAC system must maintain comfortable room temperature despite rapidly changing outdoor conditions.

    (a) Identify which fuzzy system component defines the linguistic rules for adjusting heating or cooling.
    (b) Explain briefly why a fuzzy controller is preferred over a classical PID controller in this case.

15. A medical diagnostic fuzzy expert system evaluates patient symptoms with incomplete data.

    (a) Name two internal evaluation measures that could help assess clustering or classification performance if labels are unavailable.
    (b) Suggest one future trend (mentioned in the text) that could enhance the adaptability of such systems.

### *3.2.10 Modern AI and Emerging Technologies*

1. Generative AI differs from traditional machine learning primarily because it:

    (a) Focuses only on classification tasks
    (b) Creates new data samples from learned patterns
    (c) Requires no training data
    (d) Uses only rule-based inference

2. In a Generative Adversarial Network (GAN), the generator's main objective is to:

    (a) Identify real samples from fake ones
    (b) Create synthetic data to fool the discriminator

(c) Minimize the number of training iterations
(d) Encode input features into quantum states

3. Diffusion Models generate data by:

   (a) Adding increasing levels of noise during inference
   (b) Iteratively denoising a noisy sample through a reverse diffusion process
   (c) Using a single forward pass through a generator
   (d) Performing hierarchical clustering of latent features

4. Which of the following best describes a key trade-off of Diffusion Models compared to GANs?

   (a) Less stable training but faster inference
   (b) More stable training but slower sampling
   (c) Faster training but lower fidelity
   (d) Lower computational cost and higher speed

5. Federated learning improves data privacy because it:

   (a) Encrypts all training data before sending it to the server
   (b) Trains models locally on devices and only shares weight updates
   (c) Requires no communication between client devices
   (d) Removes the need for any central aggregation

6. The **FedAvg** algorithm in federated learning is responsible for:

   (a) Splitting datasets into labeled and unlabeled subsets
   (b) Combining client updates into a single global model
   (c) Encrypting communication between clients
   (d) Detecting adversarial attacks on edge devices

7. Explainable AI (XAI) aims to:

   (a) Increase model complexity for better accuracy
   (b) Provide human-understandable reasons for AI predictions
   (c) Train models without data
   (d) Eliminate the need for feature selection

8. Which of the following is a **post-hoc** XAI method?

   (a) Decision tree models
   (b) SHAP (Shapley Additive Explanations)
   (c) Linear regression with interpretable coefficients
   (d) k-Nearest Neighbors with distance weighting

9. Neuromorphic computing differs from traditional von Neumann architectures by:

   (a) Using sequential instruction execution only
   (b) Mimicking brain-like event-driven processing and synaptic plasticity
   (c) Requiring quantum entanglement for computation
   (d) Eliminating the need for hardware components

10. Quantum Machine Learning gains potential computational speedups by exploiting:
    (a) GPU parallelism alone
    (b) Biological synapse replication
    (c) Quantum phenomena such as superposition and entanglement
    (d) Weight sharing across neural layers
11. In GAN training, the ________ network attempts to distinguish real data from generated data, while the ________ network tries to produce data that can fool it.
12. The ________ mechanism in Transformer models allows simultaneous focus on multiple input relationships, enabling parallel sequence processing.
13. Federated learning typically operates in two main settings: **cross-device**, involving numerous edge devices, and ________, which uses a small number of stable servers or data centers.
14. A hospital network wants to use sensitive patient data from multiple clinics to train a predictive model while remaining GDPR-compliant.
    (a) Which AI technique is most suitable?
    (b) Mention one security enhancement that can further protect privacy during training.
15. An AI team needs to analyze extremely high-dimensional genomic data for drug discovery, where classical algorithms struggle with the curse of dimensionality.
    (a) Which emerging AI technology offers quantum acceleration for this task?
    (b) Name one quantum property that enables this computational advantage.

MIX
Papier aus verantwortungsvollen Quellen
Paper from responsible sources
FSC® C105338

If you have any concerns about our products,
you can contact us on
**ProductSafety@springernature.com**

In case Publisher is established outside the EU,
the EU authorized representative is:
**Springer Nature Customer Service Center GmbH**
**Europaplatz 3, 69115 Heidelberg, Germany**

Printed by Libri Plureos GmbH
in Hamburg, Germany